AXOLOTL
Care Unveiled

Step-by-Step Instructions for
Creating the Perfect Axolotl Habitat

Lisa Bundy

Copyright Notice:

© 2024 [Lisa Bundy]. All rights reserved.

No part of this book may be reproduced or transmitted in any form without written permission from the publisher, except for brief quotes in reviews. For permissions, contact:

Disclaimer:

This book provides educational information and is not a substitute for professional veterinary advice. Always consult a qualified veterinarian for any health issues concerning your pet. The author and publisher are not liable for any damages arising from the use of this book. Use the information at your own risk.

TABLE OF CONTENTS

Table of Contents   3

Introduction   8

  Welcome to the World of Axolotls   9
    The Fascinating History of Axolotls   9
    Unique Features   13
    Why Axolotls Make Unique Pets   18

  Considerations Before Getting an Axolotl   22
    Legal and Ethical Considerations   22
    Laws and regulations   26
    Ethical Pet Ownership   32
    An Overview of This Guide   37

Understanding Axolotls   41

  Biology and Anatomy of Axolotls   41
    Physical Characteristics   41
    Natural Habitat and Behavior   45

  Axolotl Lifespan and Development Stages   50
    From Egg to Adult   50

Life Cycle and Growth Patterns	56
Preparing for Your Axolotl	**61**
Choosing the Right Axolotl	**61**
Where to Buy: Breeders vs. Pet Stores	61
Health Indicators to Look For	65
Essential Supplies and Equipment	**69**
Tank Size and Setup	69
Substrate and decorations	73
Filtration and Aeration Systems	77
Lighting and Temperature Control	81
Setting Up the Perfect Habitat	**86**
Creating an Ideal Environment	**86**
Water Quality: Parameters and Testing	86
Cycling the Tank: A Step-by-Step Guide	90
- Plants and Hides: Creating a Stimulating Habitat	94
Maintaining Water Quality	**98**
Regular Water Change	98
Handling Common Water Issues (Ammonia, Nitrite, Nitrate)	102
Feeding Your Axolotl	***107***
Nutritional Needs	**107**

Natural Diet in the Wild ... 107
The best foods for captive axolotls ... 111

Feeding Practices ... 115
How Often to Feed ... 115
Portion size and feeding techniques ... 119
Live Food vs. Prepared Food ... 123

Health and Wellness ... 128

Recognizing Common Health Issues ... 128
Signs of a Healthy Axolotl ... 128
Common Diseases and Symptoms ... 133

Preventative Care ... 137
Quarantine Procedures ... 137
How To Tell If Your Axolotl Is Sick ... 141
Regular Health Checks ... 146

Treating Illnesses ... 151
First Aid and Emergency Care ... 151
10 Things You Should Never Do to Axolotl ... 156
When to See a Veterinarian ... 160

Behavior and Interaction ... 165

Understanding Axolotl Behavior ... 165
Normal vs. Abnormal Behavior ... 165

Handling and Interaction Tips	170
Enrichment Activities	**174**
Toys and Enrichment Ideas	174
Creating a stimulating environment	178
Breeding Axolotls	*183*
Breeding Basics	**183**
Understanding Breeding Behavior	183
Preparing for Breeding	188
Raising Axolotl Larvae	**193**
Caring for Eggs and Hatchlings	193
Larval Development and Care	198
Advanced Care Techniques	*203*
Aquascaping for Axolotls	**203**
Designing a Naturalistic Tank	203
Advanced Plant Care	208
DIY Projects for Axolotl Enthusiasts	**212**
Building Custom Hides and Décor	212
How To Construct Custom Hides	217
How To Build Floating Islands Or Platforms	222
DIY Filter Systems	227
Axolotls in the Community	*232*

Joining the Axolotl Community	**232**
Online Forums and Groups	232
Attending Axolotl Expos and Events	241
Conservation Efforts	**246**
Supporting wild Axolotl populations	246
How to Get Involved	250
Conclusion	256
Final Tips for Successful Axolotl Keeping	**256**
Recap of Key Points	256
Appendices	**265**
Glossary of Terms	265
Troubleshooting Guide	275
References	280

INTRODUCTION

WELCOME TO THE WORLD OF AXOLOTLS

THE FASCINATING HISTORY OF AXOLOTLS

Axolotls, commonly known as Mexican walking fish, are one of nature's most fascinating species, enthralling scientists and pet owners. Their origins begin in the ancient lakes of Xochimilco and Chalco in the Valley of Mexico, where they have been a part of local tradition and a symbol of the region's unique biodiversity for ages. These lakes, relics of an extensive water system that once supported the Aztec empire, provided a perfect environment for axolotls, thanks to their calm, clear waters and plentiful vegetation.

The Aztecs, who formed their empire in the Valley of Mexico in the 14th century, were the first to document the axolotl, which played an essential role in their mythology. They thought the axolotl was the aquatic form of Xolotl, the god of fire and lightning, and the twin brother of Quetzalcoatl, the feathered serpent god. According to tradition, Xolotl turned into an axolotl to avoid being sacrificed, and his appearance in the lakes was interpreted as a celestial apparition. This mythological association gave the axolotl a mysterious air, and it became an essential aspect of Aztec society.

The axolotl's unusual traits have traditionally distinguished it from other amphibians. Unlike most salamanders, which metamorphose and adapt to a terrestrial habitat, axolotls retain their larval characteristics throughout their lifetimes, known as neoteny. This means they stay aquatic, with external gills and a finned tail, and do not develop lungs or migrate to land. This characteristic has captivated biologists by providing insights into developmental biology and evolution.

European scientists first encountered axolotls in the nineteenth century when explorers and naturalists returned them to Europe. These odd organisms immediately became a source of scientific interest. Axolotls made news in the scientific community in 1863 after Auguste Duméril, a French biologist, successfully raised them in captivity in Paris' Jardin des Plantes. Duméril's research established the axolotl as an essential model organism for studying regeneration, development, and genetics.

Axolotls have a fantastic ability to regenerate lost body parts like limbs, spinal cords, hearts, and even sections of their brains. Their regenerative ability has made them a focus of scientific investigation, with implications for human medicine and tissue engineering. Researchers seek to learn more about axolotl regeneration to improve treatments for human injuries and degenerative disorders.

Regardless of their scientific importance, axolotls confront various obstacles in the environment. The lakes of Xochimilco and Chalco have suffered from development, pollution, and the introduction of alien species, resulting in a significant drop in axolotl populations. Conservation initiatives are undertaken to protect their native habitat and avoid extinction. Local communities and international groups are working together to restore the canals of Xochimilco, eradicate invasive species, and raise awareness about the necessity of maintaining this unique species.

Axolotls have become popular pets in recent years thanks to their unique look and low maintenance requirements. As the demand for captive-bred axolotls grows, additional challenges have emerged. Responsible breeding procedures and education on animal care are critical to ensuring that axolotls survive in captivity and the wild.

Axolotls have also appeared in popular culture, including video games, cartoons, and environmental campaigns. Their unique appearance and exceptional skills make them an obvious choice for grabbing the public's imagination. This visibility has increased awareness of their conservation status and the importance of protecting their natural environment.

The history of axolotls demonstrates the interdependence of culture, science, and conservation. From their cherished place in Aztec mythology to their significance as a model organism in modern study, axolotls have impacted human understanding and imagination. Their narrative is one of resilience and adaptability, which reflects the more significant issues that biodiversity faces in our increasingly changing world.

As we continue our research and care for these beautiful animals, we are reminded of the delicate balance between the meddling of humans and the natural environment. The extraordinary trip that the axolotl has taken from ancient lakes in Mexico to contemporary laboratories and living rooms worldwide serves as a potent reminder of the significance of protecting our natural heritage for the sake of future generations. Protecting and gaining knowledge about the axolotl enables us to better understand the vast diversity and potential of life without compromising the preservation of rare and endangered species.

UNIQUE FEATURES

Axolotls, often known as Mexican walking fish, are unique organisms that pique the interest of scientists and enthusiasts alike. Their distinguishing characteristics separate them from other frogs, making them fascinating subjects of research and cherished pets. Understanding these distinguishing qualities can increase your admiration for these magnificent creatures.

One of the most distinguishing characteristics of axolotls is their capacity to retain larval qualities throughout their lives, a process known as neoteny. Unlike most amphibians that undergo metamorphosis, axolotls stay aquatic and gilled even as adults. This means they never convert from water-dwelling larvae to land-dwelling adults, remaining in their juvenile state indefinitely. Their external gills, which resemble feathery branches, protrude from both sides of their heads, lending them a peculiar and almost magical appearance. These gills are physically appealing and reasonably practical, helping axolotls collect oxygen from water efficiently.

Axolotls are best known for their regeneration ability. They can regenerate lost limbs, tail segments, and even sections of their heart, spinal cord, and brain with incredible precision. This healing ability is not only essential for survival in the wild, where injuries are common, but it also provides significant insights into the field of regenerative medicine. Researchers conduct considerable research on axolotls to better understand the mechanisms underlying their healing activities and discover possible uses in human health.

Axolotls are also distinguished by their diverse colors and patterns, resulting from genetic mutations. Wild-type axolotls are often black and speckled, which helps them blend in with their surroundings. Captive mating has resulted in several color variants, including leucistic (pale pink with red gills), albino (white with red eyes), golden albino (yellowish with red eyes), and melanoid (uniformly dark without iridophores). These many colorations add to the attractiveness of keeping axolotls as pets since owners can select from a vast range of visually stunning people.

Axolotls have a unique method of locomotion in which they walk around the bottom of their aquatic environment using their solid and fully-developed limbs. Despite being able to swim, they frequently choose to meander leisurely, earning them the nickname "Mexican Walking Fish." Their well-developed, somewhat enlarged limbs aid this activity, which is more robust than other larval amphibians. Their movements are both captivating and charming, which adds to their popularity among pet owners.

In addition to their morphological characteristics, axolotls have unique behavioral qualities. They are typically solitary creatures but can engage in various activities with their surroundings and potential partners. Axolotls are noted for their inquisitive attitude since they frequently explore their environment and investigate new objects placed into their habitat. This curiosity and their relatively peaceful nature make them exciting pets to watch.

Axolotls' eating is an intriguing component of their biology. In the wild, opportunistic carnivores eat small fish, insects, and other aquatic invertebrates. In captivity, their diet usually consists of high-quality pellets, earthworms, bloodworms, and brine shrimp. It is fascinating to see their ravenous desire and unusual eating routines, such as performing a rapid gulping motion to take in food. The owners can provide the best possible care for their animals by first understanding their nutritional requirements. Proper nutrition is essential for their health.

Axolotls have flourished in the unusual habitat of the ancient lake complex of Xochimilco, near Mexico City. This ecosystem, which consists of a network of canals and remnant lakes, is suitable for axolotls due to its cold, steady water temperatures and abundance of aquatic plants. Unfortunately, urbanization, pollution, and invasive species threaten axolotls' natural habitat, making conservation efforts crucial. Understanding their habitat requirements and risks emphasizes the importance of conservation actions.

Axolotls have a unique set of sensory adaptations. Their lateral line system, which consists of sensory organs on the sides of their bodies, allows them to perceive vibrations and movements in the water. This mechanism enables them to explore their surroundings, find prey, and avoid potential hazards. Furthermore, their strong sense of smell helps them discover food and recognize other axolotls, especially during mating season.

In scientific research, axolotls are essential models for researching developmental biology and genetic regulation. Their giant embryos and translucent skin during early development make it easy for scientists to watch and alter their growth processes. This research has shed light on vertebrate development and the possibility of regenerative therapy in humans.

Axolotls are impressive organisms with unique properties ranging from neoteny and regenerative ability to varied color morphs and behavioral peculiarities. Whether you are a pet owner or a researcher, understanding these features will improve your connection with and enjoyment of axolotls. Their unique biology continues to inspire scientific research and capture the hearts of enthusiasts worldwide.

WHY AXOLOTLS MAKE UNIQUE PETS

With their unearthly appearance and unusual mannerisms, Axolotls have captivated the hearts of pet lovers worldwide. Their particular charm stems from their uncommon appearance and the intriguing way they interact with their surroundings and those who care for them. These aquatic critters, native to Mexico's ancient lakes, convey a sense of natural wonder into households, making them unique among exotic pets.

One of the most distinguishing characteristics of axolotls is their persistent larval stage. Unlike most amphibians, which transform into new forms, axolotls maintain their juvenile features throughout their lives. This condition, known as neoteny, means they retain their external gills and aquatic lifestyle, giving them the appearance of fantastical creatures. Their feathery gills and broad, expressive faces, complete with tiny, lidless eyes, give them a permanent air of surprise or curiosity, which many people find appealing.

Caring for axolotls is a joyful experience that necessitates understanding their unique requirements. They flourish in a well-maintained aquatic environment resembling their natural home's chilly, clear waters. Setting up an axolotl tank is a fun endeavor that entails choosing the perfect substrate, ensuring proper filtration, and maintaining ideal water temps. Axolotls enjoy quiet, darkly light environments, which contributes to the peaceful aura they create in a space.

Axolotls are also fascinating pets to feed. Worms, small fish, and specialty pellets comprise most of their food. Axolotls consume with a quick suction action, exhibiting their predatory tendencies. Their feeding habits provide insight into their natural activities and enable owners to interact with them regularly.

Axolotls are recognized for their incredible regeneration abilities, which never cease to astonish. They can regenerate complete limbs, the spinal cord, heart tissue, and even sections of their brain. This extraordinary talent not only captivates pet owners but is also the topic of scientific investigation, with potential applications in regenerative medicine. Owning an axolotl provides a rare opportunity to see this fascinating process firsthand, adding a sense of wonder to their care.

Despite their unusual appearance, axolotls are very low-maintenance pets once their habitat is appropriately set up. They do not require much connection or stimulation, making them ideal for folks who lead busy lives or prefer to observe their pets rather than handle them constantly. However, they enjoy a well-decorated tank with plenty of hiding places, which may be constructed with aquatic plants and rocks, as well as specifically built axolotl hides. This keeps people healthy and stress-free and improves the aesthetic attractiveness of their surroundings.

Axolotls are also ideal for educational purposes. Their unique biology and care requirements provide an opportunity to learn about amphibian life cycles, aquatic ecosystems, and water chemistry fundamentals firsthand. This makes them especially popular in educational settings, where they can be used to teach students about biology and environmental science practically and excitingly.

Axolotls and their owners can create a delicate but powerful link. While axolotls may not show affection like dogs or cats do, they can know their owners and may react to their presence, particularly at feeding time. This recognition fosters a bond that grows with time, rewarding patience and consistency with glimpses into their eccentric personalities.

Axolotls are good conversation starters, eliciting curiosity and intrigue from friends and visitors. Their distinctive appearance and curious actions frequently prompt inquiries and discussions, making them a fascinating and one-of-a-kind addition to any household. Sharing the delight of axolotl ownership can also increase awareness about the species' conservation status and the work required to conserve their natural environment.

Axolotls are the ideal exotic pet because they combine beauty, mystery, and educational value. Their peculiar appearance and mannerisms make them entertaining, while their care requirements present a gratifying challenge for dedicated pet owners. Axolotls, whether kept for their aesthetic appeal, fascinating biology, or simply the pleasure of owning a one-of-a-kind pet, add a sense of wonder and natural beauty to our lives. Caring for an axolotl is a journey of discovery and adoration, making them an excellent choice for anybody wishing to add a touch of the remarkable to their daily lives.

CONSIDERATIONS BEFORE GETTING AN AXOLOTL

LEGAL AND ETHICAL CONSIDERATIONS

Before bringing an axolotl into your home, you should evaluate the legal and ethical implications of having this unusual and sensitive amphibian. Understanding these factors will allow you to make a reasonable and educated decision respecting animal and local regulations.

First, understand the legal criteria for axolotl ownership. Axolotls face tight control in some areas due to their position as endangered in the wild. Certain states in the United States, including California and New Jersey, have regulations that make it illegal to own axolotls without specific authorization. These regulations are intended to safeguard local ecosystems from potential incursions by non-native species while ensuring wildlife populations' conservation. Before purchasing an axolotl, study your local rules to see if ownership is permitted and if any additional licenses are required. Ignoring these restrictions may result in legal sanctions and potentially harm native wildlife.

Ethical considerations are also essential in axolotl ownership. When not managed correctly, the pet trade can contribute to the loss of natural populations. As a result, it is critical to get your axolotl from trustworthy breeders who value animal welfare and use ethical breeding procedures. These breeders ensure that axolotls are produced in captivity rather than removed from their natural habitats, thereby protecting wild populations. To help you make a responsible decision, ask potential breeders about their procedures, including their axolotls' health and living conditions.

Once you've checked the legality and identified a trustworthy breeder, consider the long-term commitment to having an axolotl. These critters can survive for more than a decade with adequate care, so preparing for this long-term commitment is critical. Evaluate your abilities to give consistent care, such as maintaining suitable water conditions, feeding a nutritious diet, and monitoring health risks. Owning an axolotl is a long-term commitment, and ensuring you can meet their needs for the rest of their lives is an important ethical consideration.

In addition to legal and ethical considerations, consider axolotl ownership's impact on your family and other pets. Axolotls require a quiet, stable habitat devoid of excessive noise and interruptions. If you have small children or other pets, consider how their presence may influence the axolotl and whether you can provide a comfortable living environment that reduces stress for all animals involved. Educate your household on axolotls' specific demands and sensitivities to build a peaceful atmosphere.

Environmental effect is another ethical factor. Axolotls require precise water conditions, such as clean, dechlorinated water and constant temperatures. This entails utilizing water conditioners, filters, and possibly aquarium chillers. Consider the environmental impact of maintaining an axolotl tank and look at ways to reduce waste and energy use. For example, employing energy-efficient equipment and reusing water as much as possible can help lessen the environmental impact of axolotl care.

Health and safety are vital. Make sure you have access to a veterinarian who has experience with amphibians. Your axolotl's well-being depends on regular check-ups and timely care for health issues. Investigate common health conditions, such as fungal infections and metabolic bone disease, and be ready to manage them if they occur. Having an emergency care plan in place shows you are committed to treating your pet ethically.

Also, the ethical concerns of breeding axolotls should be examined. If you intend to breed them, understand the genetics, health concerns, and obligations of raising and rehoming offspring. Uninformed breeding can result in genetic problems and contribute to overpopulation in the pet industry. Responsible breeding practices involve preserving genetic diversity, avoiding inbreeding, and ensuring that offspring are raised in competent and prepared homes.

Consider the overall impact of axolotl ownership on conservation efforts. Supporting organizations dedicated to conserving wild axolotl populations and natural habitats can help them survive in the wild. Educating others about the importance of conservation and the difficulties encountered by wild axolotls can increase awareness and support for these efforts.

Maintaining an axolotl requires careful consideration of legal, ethical, and practical concerns. By thoroughly researching and planning for the responsibilities involved, you may ensure that your decision benefits both the species and the more considerable conservation efforts. Respect for axolotl ownership's legal and ethical aspects demonstrates a dedication to careful and humane pet care.

LAWS AND REGULATIONS

When considering adding an axolotl to your household, it's

critical to understand the complex network of laws and regulations governing these distinctive amphibians' ownership. These legal considerations are intended to safeguard both the species and the local ecosystem, ensuring that axolotls are managed properly and ethically.

Axolotls are native to Mexico's lakes, Xochimilco and Chalco, and their wild populations are critically endangered owing to habitat destruction, pollution, and invasive species. As a result, international, national, and municipal legislation have been enacted to save this species. Understanding these laws is critical for every potential axolotl owner.

Axolotls are included in Appendix II of the Convention on International Trade in Endangered Wild Fauna and Flora (CITES). This status indicates that, while trade is not forbidden, it is regulated to ensure that it does not jeopardize the species' survival in the wild. If you want to import or export axolotls, you must follow CITES regulations, which involve permits and precise requirements to ensure that the creatures are dealt with sustainably and ethically.

In the United States, axolotl ownership rules vary significantly from state to state. For example, owning an axolotl requires specific authorization in California and New Jersey. These states have severe legislation to preserve native ecosystems from potential invasions by non-native species, which could happen if axolotls are released into nature. Violations of these restrictions may result in heavy fines and animal confiscation. In contrast, states like Texas and Florida have more lax rules, permitting axolotl possession without special permission, while local governments may have extra requirements.

To negotiate these regulations, consult a local wildlife agency or your state's Department of Fish and Wildlife. They can supply the most current and accurate information about the legal status of axolotl ownership in your area. Talking with local pet businesses or breeders familiar with state rules can provide helpful insights and guidance.

In Canada, axolotl ownership rules vary by province. For example, in British Columbia, axolotls are lawful to own but must not be released into the wild to conserve native species and ecosystems. In Ontario, individual communities may have restrictions, so check with local authorities before purchasing an axolotl.

Axolotl ownership is generally permitted in the United Kingdom. However, owners must follow the Animal Welfare Act 2006, which requires pets to be kept in a proper environment, with nutrition, and with care. This act emphasizes pet owners' responsibility for the welfare of their animals, especially axolotls. To safeguard local ecosystems, the Wildlife and Countryside Act of 1981 restricts the release of non-native animals into the wild, including axolotls.

Regulations in European Union countries can differ significantly. In Germany, axolotl ownership is legal, but rigorous animal welfare laws require owners to provide adequate care and living circumstances. Some nations, such as Norway, prohibit the ownership of axolotls because of worries about invasive species and environmental damage. Consulting with local authorities or wildlife agencies is critical to ensuring compliance with national and local regulations.

Australia's rules are incredibly stringent due to the

country's unique and endangered ecosystems. The Environment Protection and Biodiversity Conservation Act of 1999 prohibits the importation of axolotls. They can own property in some states and territories, such as Victoria and New South Wales if the necessary permits are obtained. Owners must follow all norms and regulations to avoid any potential environmental damage.

Understanding and adhering to these laws is both a legal and ethical responsibility. By following restrictions, axolotl owners help to conserve the species and protect local ecosystems. Furthermore, knowing and implementing these regulations helps to minimize the spread of misinformation and promotes responsible pet ownership in the community.

Beyond the legality of ownership, ethical concerns are critical in maintaining an axolotl. It is vital to source your axolotl from trustworthy breeders who use ethical breeding procedures. These breeders prioritize their animals' health and well-being, ensuring they are raised in appropriate surroundings and free of genetic abnormalities. Avoiding breeders who engage in unethical behavior, such as inbreeding or lousy living conditions, benefits the overall health of the captive axolotl population.

Furthermore, ethical ownership means never releasing axolotls into the wild. Such actions can have disastrous consequences for local ecosystems, including the extinction of native species and the spreading of diseases. Educating others on the value of responsible pet ownership and axolotls' unique needs can also promote ethical practices in the community.

Those considering breeding axolotls must comprehend the legal and ethical consequences. Breeding should be approached with care, considering the genetic variety and the health of both parents and offspring. To avoid overcrowding and neglect, ensuring that the kids are in demand and placed in competent and prepared households is critical.

Conservation activities can help axolotls survive in the wild. Supporting groups that work to restore habitat, limit pollution, and safeguard wild species can significantly influence. Participating in citizen science programs, such as monitoring local rivers for amphibian numbers, can help achieve larger conservation goals.

The decision to possess an axolotl requires navigating a complicated environment of laws and regulations that differ by country, state, and municipality. Understanding and adhering to these standards demonstrates that you are a responsible and ethical pet owner. By purchasing axolotls from trustworthy breeders, following legal regulations, and supporting conservation efforts, you help ensure the well-being of these beautiful creatures and protect our shared environment.

ETHICAL PET OWNERSHIP

Ethical pet ownership entails a wide range of obligations, including understanding the animal's needs, complying with legal requirements, and encouraging the well-being of the pet and the larger ecosystem. Owning an axolotl, a unique and delicate amphibian, necessitates paying close attention to these details to ensure you provide the best possible care while adhering to environmental and regulatory restrictions.

When introducing an axolotl into your house, the first step

is to learn about their specific needs. Axolotls require a calm, stable aquatic environment with clean water, an adequate diet, and little stress. This includes establishing a proper tank with insufficient filtration and temperature control and supplying various meals that fit their nutritional requirements. By committing to learning about their care requirements, you may help your pet live a healthy and enjoyable life.

Another essential part of responsible ownership is ethical sourcing for your axolotl. This entails selecting reliable breeders who value their animals' health and genetic diversity. Reputable breeders guarantee that axolotls are bred under proper conditions and not stolen from the wild, where they are critically endangered. Avoiding black market sellers and unethical breeding practices promotes sustainable and humane pet trade practices.

Legal considerations are also crucial while owning an axolotl. Unique rules and regulations in different regions govern axolotl ownership, importation, and breeding. Owning an axolotl is forbidden in California and New Jersey without specific licenses because of worries about invasive species and environmental damage. Other states, such as Texas, have more lax regulations. It is critical to consult with local wildlife agencies or regulatory bodies to ensure you follow all legal regulations. Ignoring these restrictions might have legal penalties and harm local ecosystems.

Ethical ownership includes establishing a setting closely resembling the axolotl's natural habitat. This includes frequently testing for dangerous compounds such as ammonia, nitrites, and nitrates and keeping the water temperature within the recommended range of 60-68 degrees Fahrenheit. Providing hiding spaces and a diversity of textures in the tank reduces stress and promotes natural behaviors. The dedication to providing an appropriate habitat demonstrates a vital concern for the animal's well-being.

Understanding and detecting indications of stress and illness in axolotls is part of responsible pet ownership. Axolotls are susceptible to various health difficulties, including fungal infections, injuries, and stress-related diseases. Regular monitoring and timely veterinarian care

are needed. Finding an expert veterinarian who works with amphibians might be difficult, but it is vital to give appropriate medical care when needed.

Another crucial responsibility is to prevent axolotls from being released into the wild. Axolotls are not native to most areas outside their natural home in Mexico, and releasing them can damage local ecosystems. They can outcompete native species for resources, spread diseases, and become invasive. Educating yourself and others on the necessity of keeping pets enclosed and never releasing them into the wild helps to preserve local biodiversity.

Breeding axolotls properly is also an essential ethical topic. To breed axolotls, you must first understand their genetics, health hazards, and the commitment required to care for the kids. Ensure that the axolotls you breed are in demand and that you can find acceptable, competent homes. Overbreeding and inadequate care for offspring can result in neglect and health problems, compromising ethical standards.

Engaging with the larger community of axolotl owners can help you care for your pet more responsibly. Joining forums, courses, and local groups allows you to learn from more experienced owners while sharing your insights. This community participation fosters a culture of responsible ownership and lifelong learning.

Supporting conservation efforts is an essential component of responsible pet ownership. Due to habitat degradation and pollution, axolotls are highly endangered in the wild. Contributing to organizations that preserve natural ecosystems and promote sustainable practices can significantly influence. Participating in citizen science programs, such as monitoring local rivers for amphibian numbers, contributes to conservation aims.

Owning an axolotl necessitates a commitment to ethical standards, ranging from sourcing and habitat management to understanding legal obligations and conservation efforts. By accepting these obligations, you can ensure that your axolotl flourishes in a healthy, supportive environment while helping preserve this beautiful species. Ethical pet ownership entails providing basic requirements and instilling a strong sense of respect and responsibility for the animal's well-being and the ecology in which it lives.

AN OVERVIEW OF THIS GUIDE

Welcome to our thorough guide to axolotl care. This guide is intended to be your go-to reference for all you need to know about growing and caring for a healthy and happy axolotl. Whether you are a first-time owner or an expert aquarist wishing to broaden your knowledge, this guide will cover all elements of axolotl care clearly and plainly.

Starting with the fundamentals, we explore the intriguing history of axolotls. These fascinating species, native to Mexico's lakes, have a rich cultural and biological heritage. Understanding their beginnings allows us to appreciate their individuality while learning about their natural habits and demands. We look at their significance in Aztec mythology and how they have intrigued scientists and enthusiasts.

As you go through the tutorial, you'll discover thorough instructions on preparing for your axolotl. Every element is addressed, from picking the perfect tank to decorating it. The initial setup is critical to your axolotl's health, and we provide step-by-step instructions to assist you in building a safe and exciting environment. This contains information about water quality, filtration systems, and maintaining the optimal temperature. Following these instructions may ensure that your axolotl lives in a pleasant and healthy environment.

Feeding your axolotl is another critical component of their care, and we provide detailed information on their nutritional requirements. To remain healthy, axolotls must meet specific dietary requirements. We review the types of food they thrive on, including both live and prepared options, and offer practical advice on feeding schedules and approaches. Understanding their eating patterns will help you keep them healthy and vital.

Health and wellness are essential components of any pet care regimen, and this book provides in-depth insights into common health issues and preventative actions. Axolotls, while resilient, are prone to certain diseases if not properly cared for. We outline both signals of excellent health and symptoms of potential problems, giving you the knowledge to spot and handle issues early on. In addition, we cover basic first aid and emergency treatment, preparing you to act quickly if necessary.

This guide also addresses crucial themes such as behavior and interaction. While axolotls are not well-known for their social behavior, recognizing their natural inclinations and habits can help you connect with them more effectively. We provide advice on observing and interpreting their actions, resulting in a more exciting and enriching environment for your pet.

Many fans are interested in breeding axolotls, and this article will provide an overview of the procedure. From recognizing breeding behaviors to caring for eggs and larvae, we provide insights into axolotl care's challenging yet gratifying element. Whether you're thinking of breeding for the first time or want to enhance your methods, these sections will help.

We go over advanced maintenance techniques such as aquascaping and D.I.Y. projects for individuals who appreciate personalizing their aquariums. Creating a stylish and practical habitat benefits your axolotl while also improving the aesthetics of your setup. We share innovative ideas and valuable tips to help you create a beautiful and healthy environment.

This page discusses the broader context of axolotl care, including conservation efforts and community engagement. Axolotls are an endangered species, and their preservation is critical. We emphasize how you can help preserve them and connect with other axolotl aficionados. By developing a feeling of community, we can all work together to ensure the survival of these incredible creatures.

Throughout this guide, you will discover a wealth of information from specialist sources and experienced keepers. We aim to offer you the tools and knowledge to care for your axolotl and ensure it lives a long, healthy, and whole life.

Chapter 1

UNDERSTANDING AXOLOTLS

BIOLOGY AND ANATOMY OF AXOLOTLS

PHYSICAL CHARACTERISTICS

Axolotls, often known as Mexican walking fish, are exciting creatures with unique morphological traits that set them apart in the animal kingdom. Their distinct appearance has fascinated scientists and pet lovers, providing infinite wonder and curiosity.

One of the most distinguishing characteristics of axolotls is their constant larval state, known as neoteny. Unlike most amphibians, which undergo metamorphosis to become adult forms, axolotls maintain their juvenile characteristics throughout their lives. This ensures they remain aquatic and retain their external gills, which gives them their distinctive frilly headgear. These gills are functional and aesthetically pleasing, with feathery appendages that can be pink, crimson, or even white depending on the axolotl's colors and health.

Axolotls come in a range of colors, each with their unique charm. Wild-type axolotls are usually dark brown or black with speckles, giving them good concealment in their natural environment. Albino axolotls, on the other hand, have entirely white bodies with pink gills and eyes, giving them a ghostly, ethereal look. Leucistic axolotls resemble people with albinism but have black eyes, whereas melanoid axolotls are entirely black, lacking the iridescent colors of other color varieties. There are also golden albinos, which have a vibrant yellow color. These color variations are the product of selective breeding, which has produced a diverse range of visually appealing axolotls for fans to enjoy.

An axolotl's body is elongated and narrow, making it ideal for an aquatic lifestyle. They can grow up to 12 inches long, but most in captivity are slightly smaller, averaging 9 to 10 inches. Their smooth and somewhat slimy skin allows them to glide effortlessly through water. This streamlined body form is enhanced by a long, finned tail extending from the torso's end, assisting with propulsion and maneuverability.

Axolotls have four short, stubby legs with sensitive fingers that they use to travel along the aquatic floor. Despite being called "walking fish," they move in a slow, methodical shuffle. Axolotls frequently utilize these legs to grip and manipulate their prey; hence, they play an essential role in feeding.

One of the most fascinating qualities of axolotls is their regenerating powers. They can regrow complete limbs, tails, and even sections of their spinal cord, heart, and brain. This ability is not just a survival mechanism but also the focus of intensive scientific inquiry, as understanding how axolotls regenerate could have significant consequences for human medicine. Observing an axolotl recover a limb demonstrates the species' incredible tenacity and adaptability.

Axolotls have a giant, flat skull and a broad mouth that often appears smiling, making them popular among pet owners. Their eyes have no lids and vary in color depending on their genetic composition. Wild-type axolotls typically have dark eyes, whereas people with albinism have pinkish-red eyes. Positioning their eyes on the sides of their heads gives them a panoramic view of their environment, which helps identify predators and prey equally.

Axolotls also have remarkable sensory talents. They have a lateral line system, a network of sensitive receptors along the sides of their bodies that allows them to perceive vibrations and movements in the water. This system is comparable to that seen in fish and is essential for navigating their surroundings, locating food, and avoiding danger.

Despite their strange look, axolotls are relatively straightforward to care for as long as their specialized requirements are satisfied. They flourish in cool, clear water with a consistent temperature and minimal levels of light. Their tanks should be large and stocked with hiding spots, as they can be bashful and appreciate having a haven.

Their nutrition is very directly related to their physical

traits. Axolotls are carnivorous, eating a range of small aquatic animals such as worms, insects, and fish. In captivity, they can be fed a combination of specially prepared pellets and live or frozen items. Watching an axolotl eat is intriguing since they employ a suction mechanism to attract their prey, demonstrating their predatory instincts.

Axolotls are truly unusual, with various characteristics distinguishing them from other amphibians. Every part of their body, from their distinctive gills and color variations to their regenerating abilities and sensory systems, recounts a story of adaptation and survival. Owning an axolotl provides a glimpse into a world of natural beauty, with each attribute playing an essential role in their incredible survival.

NATURAL HABITAT AND BEHAVIOR

Axolotls, also known as Mexican walking fish, are native to the ancient and once-vast lakes of Xochimilco and Chalco in the Valley of Mexico. These lakes, remnants of a vast network surrounding the Aztec city of Tenochtitlan, created a unique and rich habitat that molded the axolotl's specific traits and behaviors.

Axolotls' natural environment is defined by the chilly, clear waters of these high-altitude lakes, which are more than 7,000 feet above sea level. The water temperature in these lakes is relatively stable, ranging from 60 to 64 degrees Fahrenheit. This chilly climate is critical to the well-being of axolotls since it affects their metabolic rates and overall health. The lakes are also shallow, with a complex network of canals, marshes, and floating gardens known as chinampas that contribute to a diversified aquatic environment.

Axolotls flourish in these calm waters, which are abundant in aquatic flora. The plants provide critical cover, breeding grounds, and a copious food source. This lush underwater ecosystem supports a diverse range of life forms, including small fish, insects, crabs, and worms, which are the primary feed of axolotls. The lush vegetation also protects axolotls from predators and provides a safe habitat for them to lay their eggs.

In their natural environment, axolotls are primarily nocturnal. They spend the day hidden amid the plants and residue at the lake's bottom, emerging only at night to forage for food. Their feeding approach is remarkable; they use their excellent sense of smell and lateral line system to detect vibrations and movements of prey. When an axolotl detects its target, it utilizes a quick suction action to engulf it, which is both efficient and indicative of its predatory character.

Axolotls exhibit a variety of behaviors that are appropriate for their aquatic existence. They are generally bottom dwellers, only moving above the surface if necessary. While not strong enough for quick swimming, their limbs are ideal for wandering along the lakeshore and navigating through the dense foliage. These limbs also aid them in foraging for food, allowing them to sift through the substrate in search of buried prey.

One of the most fascinating habits of axolotls is their reproductive cycle. Breeding usually happens in the spring, spurred by changes in water temperature and sunshine. During courtship, males and females perform a beautiful dance in which the male lays spermatophores that the female gathers to fertilize her eggs. The female deposits her eggs singly and attaches them to plants and other surfaces. This strategy keeps the eggs securely buried and safe from predators.

Axolotls are also noted for their capacity to regrow lost body parts, both a natural and a biological marvel. This regenerative ability enables them to recover from injuries that would kill other organisms, enabling their survival in the harsh and unpredictable environment of their native lakes. Seeing an axolotl recover a lost limb provides insight into their tenacity and adaptability.

Axolotls' natural environment, however, has suffered significant problems throughout time. Urbanization, pollution, and the introduction of exotic species have significantly affected the size and quality of native lakes.

The Xochimilco canals, which contain most of the remaining wild axolotls, are now a UNESCO World Heritage site, but they are still threatened by environmental degradation. Conservation initiatives are underway to restore and conserve this crucial ecosystem, ensuring axolotls can thrive in their native setting.

Despite these hurdles, axolotls have evolved to live in various habitats, including captivity. When kept as pets, their tanks are designed to mimic the chilly, plant-rich waters of their natural lakes. Providing a similar environment in captivity promotes their health and natural behaviors. Owners frequently adorn tanks with plants, pebbles, and hiding spots to mimic the axolotls' natural habitat, promoting security and well-being.

Axolotls' natural habitats and behaviors demonstrate their adaptability and resilience. These fascinating species have evolved to flourish in a distinct and challenging environment, acquiring habits and characteristics that assure survival. Understanding their natural habitats and behaviors sheds light on their maintenance and conservation, emphasizing the necessity of maintaining these historic lakes for future generations. Axolotls are more than simply intriguing pets; they also represent a fragile and intricate ecology that demands our preservation and respect.

AXOLOTL LIFESPAN AND DEVELOPMENT STAGES

FROM EGG TO ADULT

The axolotl's life cycle is a fascinating trip demonstrating the wonders of biology and development. From the moment an egg is fertilized to the full-grown adult stage, this fascinating frog undergoes intricate modifications that highlight its perseverance and distinctiveness.

It starts with breeding, usually triggered by environmental cues like temperature and light fluctuations. In the wild, this typically occurs in the spring when conditions are ideal. The male axolotl initiates courtship by dancing, demonstrating his agility and strength to attract a female.

He places packets of sperm, known as spermatophores, onto the substrate. The female then collects these packets with her cloaca to fertilize her eggs internally.

Following fertilization, the female deposits her eggs singly, attaching them to plants or other sturdy objects in the water. Each egg is coated in a jelly-like material that provides protection and stability. An axolotl can lay between 100 and 1,000 eggs in a single reproductive episode, ensuring that even if many do not survive, enough will reach adulthood to sustain the species.

Axolotl development begins in these tiny eggs, which measure 1 to 2 millimeters in diameter. Initially, the eggs look like little dark spots, but the embryos develop within them as time passes. This stage is significant because embryos begin to produce essential structures, such as the neural tube, which will later develop into the spinal cord and brain. Depending on the water temperature, hatching takes two to three weeks. Warmer water accelerates development, whereas cooler water slows it down.

Axolotls hatch and enter the larval stage. These larvae are tiny, measuring only a few millimeters long, and aquatic. They have gills for breathing underwater and a rudimentary digestive system appropriate for their initial meal of microscopic creatures. At this point, they rely primarily on their yolk sacs for nourishment until they can seek food independently. Their gills are conspicuous and feathery, spreading from the sides of their heads, allowing for efficient oxygen exchange in the water.

As they grow, axolotls develop increasingly distinct traits. Within a few days or weeks, they begin grazing on small live meals such as brine shrimp, daphnia, and micro worms. This food is high in protein, which is required for quick growth and development. During this time, they have a ravenous appetite and eat regularly to fuel their growth.

An exciting component of axolotl development is the retention of larval characteristics throughout adulthood, called neoteny. Unlike most amphibians, which undergo complete metamorphosis, axolotls do not transform into terrestrial forms. They stay utterly aquatic throughout their lifetimes, with gills, finned tails, and juvenile body form. For years, scientists and biologists have been interested in this uncommon characteristic, which provides insights into evolutionary biology and developmental processes.

Over several months, axolotls grow in size and complexity. They resemble miniature copies of their adult counterparts by reaching one to two inches in length. Their limbs become increasingly developed, with tiny fingers and toes that help them negotiate the bottom of their watery environment. Initially delicate and small, their gills create more extensive branching, increasing their breathing efficiency.

Axolotls reach sexual maturity between six months and a year of age. They usually measure between six and ten inches in length at this time, though they can develop to be as long as twelve inches. Males and females have distinct sexual maturation characteristics. Males have longer, more elongated bodies and more significant, visible cloacal regions holding their reproductive organs. Females, on the other hand, have rounder and plumper bodies, particularly while carrying eggs.

Despite attaining sexual maturity, axolotls maintain their larval traits, such as gills and finned tails. This everlasting juvenile state allows them to remain aquatic, unlike other amphibians that shift to land-based life stages. This adaptation has proven useful in their natural habitat, where the watery environment of the Xochimilco and Chalco lakes provides adequate supplies and protection.

Throughout their life, axolotls demonstrate incredible healing ability. They can regrow missing limbs, tails, and even segments of their spinal cord, heart, and brain. This regenerating capacity is an intriguing biological phenomenon and an important survival mechanism. In the wild, predators or environmental risks can cause serious injury or death. Conversely, Axolotls may regenerate injured or lost body parts, allowing them to recover from injuries and continue to thrive.

Axolotls in captivity can survive for 15 years or more with adequate care. Providing a stable environment with clean, cool water, a balanced diet, and low stress is critical for their longevity. Regularly monitoring water quality and temperature and offering hiding places and a diverse diet help improve their overall health and well-being.

Observing the development of axolotls from egg to adult provides insight into the complexity of amphibian biology and the miracles of natural selection. Their path from tiny, delicate embryos to fully formed, neotenic adults exemplifies life's perseverance and diversity. Whether in their natural habitats or captivity, axolotls continue to captivate and inspire, reminding us of the natural world's tremendous diversity and adaptation.

Axolotls traverse their aquatic habitat with a combination of sensitivity and strength, symbolizing the delicate balance of life. Their ability to survive in a specialized ecological niche and their distinct developmental characteristics highlight the need to protect their natural environments. Conservation efforts for wild axolotl populations are critical for their survival and the ongoing study and appreciation of these unique species.

Finally, the story of an axolotl's existence is one of constant growth and adaptation. From the early stages of development within an egg to the fully developed, regenerating adult, each step presents distinct obstacles and astounding developments. Understanding and appreciating these stages offers vital insights into the larger fabric of life, emphasizing the interconnection of all living things and the settings in which they live. Axolotls teach us about nature's wonders and the intricate mechanisms that keep life going throughout their life cycle, providing infinite possibilities for learning and discovery.

LIFE CYCLE AND GROWTH PATTERNS

Axolotls, with their unique life cycle and growth patterns, are a fascinating example of amphibian development that continues to captivate scientists and enthusiasts alike. The life of an axolotl, from fertilized egg to fully grown adult, is a story of tenacity, adaptability, and distinct biological processes.

It all starts with laying eggs, a critical stage in the axolotl's life cycle. Following successful courting and fertilization, female axolotls lay hundreds of eggs in clusters. These eggs, covered in protective jelly, are adhered to underwater plants and other surfaces, ensuring their survival in the aquatic environment. The tiny embryos inside the eggs are initially simple, but they quickly develop the key characteristics that distinguish axolotls.

As the embryos develop, they begin to take on more recognizable shapes. The eggs hatch, and larvae emerge after a few weeks, depending on the temperature of the water. These newly hatched larvae are tiny and delicate, measuring only a few millimeters long. They quickly adapt to their aquatic environment, having external gills that allow them to breathe underwater and a primitive digestive system capable of processing microscopic food particles.

During the early larval stage, axolotls rely exclusively on their yolk sacs for food. This stage is crucial for their development because they grow swiftly and begin to develop the characteristics required for independent survival. Within days or weeks, they start grazing on minute animals like protozoa and small invertebrates, moving from yolk reliance to active foraging.

Axolotls' morphological characteristics alter significantly as they develop. Initially small and basic, their gills grew increasingly intricate and branching, increasing their ability to take oxygen from the water. Their limbs begin as tiny buds and progressively develop into digits, allowing them to travel more effectively along the lakebed. The tail, with its unique fin, becomes longer and more muscular, enhancing propulsion and agility.

One of the most fascinating elements of axolotl development is its neotenic behavior. Unlike most amphibians, which undergo complete metamorphosis before transitioning to a terrestrial habitat, axolotls retain their larval features throughout their lives. This means they stay aquatic, with external gills and a finned tail, and do not develop lungs or migrate to land. This distinguishing feature provides significant evolutionary benefits, allowing axolotls to exploit a stable aquatic niche without drastic environmental changes.

Axolotls are considered sexually mature between the ages of six months and a year. They usually are six to ten inches long at this point, but they can grow to twelve inches or longer later in life. Sexual maturity causes visible changes, especially in the development of reproductive organs. Males have a longer body and a larger cloacal area, whilst females are often rounder and more robust, particularly when carrying eggs.

Despite their age, axolotls retain their larval characteristics, such as unique gills and an aquatic existence. This neotenic trait enables them to stay in their chosen habitat, the calm, clean waters of high-altitude lakes. They continue to thrive in this environment, eating small fish, worms, and other aquatic organisms. Their growth slows but does not stop, with some people progressively increasing in size over time.

Axolotls' regenerating powers add to the intricacy of their life cycle. These creatures can regenerate missing limbs, tails, and organs like the heart and brain. This extraordinary capacity allows them to heal from injuries that would kill other animals, increasing their chances of survival in the wild.

Throughout their life, axolotls demonstrate predatory efficiency and delicate curiosity. Their broad, expressive faces and seemingly constant smiles conceal a ravenous appetite for prey. Observing an axolotl hunting demonstrates its versatility and finely tuned sensory systems, which include a lateral line system for detecting vibrations and movements in the water.

Axolotls can live for over a decade, with some surviving 15 years or more in captivity under ideal conditions. Maintaining their health and well-being necessitates a consistent environment with clean, cool water, adequate nutrition, and low stress. Regular monitoring and care ensure these exciting creatures flourish and expand their lives.

Axolotls' life cycle and growth patterns demonstrate these unique amphibians' extraordinary adaptability and durability. From their start as tiny, fragile larvae to their fully grown, neotenic adult forms, each stage of their life is defined by substantial modifications and adaptations that allow them to thrive in their aquatic environments. Their story is one of constant growth and regeneration, a living witness to the miracles of natural evolution and the complex processes that keep life going.

Chapter 2

PREPARING FOR YOUR AXOLOTL

CHOOSING THE RIGHT AXOLOTL

WHERE TO BUY: BREEDERS VS. PET STORES

When opting to welcome an axolotl into your life, one of the first and most crucial decisions you'll have to make is where to get your new aquatic companion. There are two primary options: buying from breeders or pet retailers, each with its benefits and downsides.

Buying an axolotl from a breeder has various advantages, especially if you want a healthy, well-cared-for animal. Breeders are usually thoroughly aware of the species and are enthusiastic about their care and well-being. They often raise axolotls in ideal conditions, ensuring that water quality, temperature, and nutrition are carefully monitored. This attention to detail usually results in healthier axolotls with fewer genetic abnormalities or diseases.

Breeders also offer a plethora of information and support, which is unique, to artist-time owners. They can provide specific advice on tank layout, feeding schedules, and overall maintenance based on their significant knowledge. This advice can help you create the most extraordinary environment for your new pet, improving the likelihood of a seamless transition and a happy, healthy axolotl.

Another advantage of buying from a breeder is the ability to select from a wide range of axolotl morphs. Breeders frequently specialize in different color variations and genetic features, allowing you to find an axolotl that meets your needs. Breeders can give a variety of alternatives, including classic wild-type axolotls, striking albinos, and unusual melanoid.

However, there are certain things to consider when buying from a breeder. Depending on their location, transportation may be required, which can be stressful for the axolotl and necessitates delicate treatment to assure their safety. Furthermore, breeders may have waiting lists, especially for unusual morphs, so you must be patient and prepare ahead.

On the other hand, purchasing an axolotl from a pet store may be more convenient, especially if you like to see the animal in person before making a purchase. Pet retailers frequently have axolotls on hand, allowing you to select one and take it home the same day. This quick availability may be enticing, especially if you want to set up your tank and begin your axolotl-keeping experience.

Pet stores are an excellent choice if you like getting all your supplies in one spot. Many places sell axolotls alongside tanks, filters, food, and other necessary equipment, making getting all you need in one trip simple. This might be especially useful for new owners still learning about Axolotl Care's unique needs.

However, being informed of the possible drawbacks of purchasing from pet retailers is critical. The circumstances in which axolotls are stored vary significantly from store to store. Some pet retailers may need more specialist knowledge to provide optimal care, resulting in axolotls being housed in unsuitable settings. Poor water quality, insufficient temperature management, and unsuitable meals impact the animals' health and well-being.

Before acquiring an axolotl from a pet store, carefully investigate the conditions in which they are housed. Look for evidence of good health, such as clear eyes, unbroken limbs, and vivid gills. Please inquire about the staff's care routines and assess their knowledge and readiness to answer inquiries. A good pet business should be able to provide basic axolotl care knowledge while demonstrating a commitment to its animals' well-being.

Consider the possibility of impulse purchases when visiting a pet store. The convenience of rapid buying can sometimes lead to rash judgments without appropriate planning. Before purchasing an axolotl, ensure you've researched and are well prepared to suit its demands.

Both breeders and pet retailers provide feasible options for obtaining an axolotl, each with its benefits and drawbacks. Breeders often provide healthier, better-cared-for animals with access to expert advice and a diverse range of variants.

Pet stores offer convenience and fast availability, but thorough evaluation is required to guarantee that the axolotls are kept in proper circumstances. Whichever choice you choose, adequate research and planning are essential for ensuring a happy and healthy environment for your new axolotl. Making an informed decision can help you get your axolotl-keeping adventure off to a good start, providing a gratifying and delightful experience for you and your new aquatic companion.

HEALTH INDICATORS TO LOOK FOR

When you adopt an axolotl, you prioritize its health and well-being. Recognizing the indications of a healthy axolotl can help you provide the best care and detect any potential problems early on. Understanding these indications is critical for ensuring your pet's health and lifespan.

A healthy axolotl has vivid and transparent exterior gills. These feathery appendages on either side of the skull are the major respiratory organs. They should be brilliantly colored and complete in a well-kept habitat, usually vibrant pink or red. The axolotl's gills should move gently with the current of the water or when it breathes. If the gills appear pale, minor, or receding, it could suggest stress or poor water quality.

A healthy axolotl's skin is also an essential indicator. It should be smooth, with no lesions or discolorations. Axolotls have a little sticky mucus layer on their skin that protects them from infections and aids in osmotic balance. Any evidence of fungal development, such as white, cotton-like patches or red, inflamed spots, may indicate a health issue that requires quick attention.

Their behavior heavily influences Axolotls' health. A healthy axolotl is generally active and curious, exploring its tank and reacting to changes in the surroundings. It should swim gracefully and smoothly, frequently walking along the tank's bottom with its limbs. Lethargy, uncontrollable floating, or excessive concealment may indicate distress or disease.

Another critical health indication is the axolotl's hunger. These critters are usually voracious eaters, gobbling food like worms, pellets, or small fish. A healthy axolotl should have a constant appetite and be interested in feeding times. A sudden loss of appetite or trouble eating may indicate underlying health conditions, such as digestive disorders or infections.

The eyes of an axolotl can reveal information about its health. They should be clear, without cloudiness or puffiness. While certain axolotl morphs have naturally distinct eye colors, any deviation from their typical appearance demands deeper examination. Swollen or clouded eyes may suggest an infection or poor water quality, requiring immediate attention.

Regular examination of the axolotl's limbs and tail is also essential. Healthy limbs are well-formed and free of abnormalities and injuries. The tail should be smooth and straight with no kinks or evidence of necrosis. Minor injuries may recover over time due to axolotls' exceptional regenerating ability, but significant damage or frequent concerns can suggest deeper problems.

The water quality heavily influences the health of the axolotl in its tank. Maintaining proper water conditions is critical for avoiding numerous health problems. Regularly test the water for ammonia, nitrite, and nitrate levels, and keep the pH level between 6.5 and 8.0. Axolotls thrive in clean, cool water between 60 and 68 degrees Fahrenheit. Poor water quality can cause stress, disease, and declining health outcomes.

The axolotl's general physical condition is another indicator of good health. It should appear full and well-fed but not obese. The axolotl's body should be rounded, indicating good feeding. Weight loss, a hollow body, and bloating are all warning signs that something is awry. A diversified diet high in protein and nutrients aids in the maintenance of their physical condition.

Interactions with other tank dwellers, if any, can reveal information on the axolotl's health. While they are often solitary creatures, they should not show signs of extreme hostility or tension toward their tank mates. Peaceful cooperation suggests a stable environment and a healthy axolotl.

Continuous monitoring of these health markers is critical for the well-being of your axolotl. You can help your pet thrive by paying attention to changes in their look, behavior, and environment. A healthy axolotl is a delight to watch, with its distinct motions and habits that provide infinite intrigue and company. Regular maintenance, such as keeping ideal water conditions and feeding a balanced diet, will help your axolotl live a long and healthy life.

ESSENTIAL SUPPLIES AND EQUIPMENT

TANK SIZE AND SETUP

Setting up the ideal tank for your axolotl is critical to maintaining their health and happiness. The procedure begins with choosing the appropriate tank size, as the available space significantly impacts your pet's well-being. A tank of at least 20 gallons is suggested for a single axolotl. This gives them room to wander about and decreases the likelihood of water quality problems. If you intend to keep more than one axolotl, you will need a larger tank to avoid overcrowding, with an extra 10 gallons for each additional axolotl.

Once you've chosen the appropriate tank, the following step is to set it up to resemble the axolotl's natural habitat. Begin with a proper substrate. Fine sand is the most excellent option since it eliminates any ingestion dangers associated with gravel. Axolotls snap at their food and may accidentally swallow pebbles, which can cause significant health problems. In contrast, fine sand is both safe and easy to clean.

After installing the substrate, consider adding some hiding locations and decorations. Axolotls are nocturnal animals that like having places to hide during the day. Aquarium-safe decorations include caves, tunnels, and P.V.C. pipes. These hiding places provide comfort and aid in relieving stress, which is essential for their general health. Plants, whether real or fake, can also make an excellent addition. They provide cover while creating a more natural and stimulating environment for your axolotl.

Another important consideration when setting up a tank is water quality. Axolotls thrive in calm, clear waters. Maintaining the proper water temperature is critical, ideally between 60 and 68 degrees Fahrenheit. Higher temperatures can be stressful and contribute to health issues. To accomplish this, you may require an aquarium chiller, especially in a warmer environment. Regularly monitoring the water temperature ensures that your axolotl is comfortable.

Filtration is vital for keeping water clean and free of pollutants. A high-quality filter that performs both mechanical and biological filtration is required. However, axolotls enjoy mild water flow, so select a filter that does not produce powerful currents. Sponge filters are frequently a good choice because they are both practical and mild to axolotls. Regular filter maintenance, including cleaning and replacing media as needed, will contribute to a healthier atmosphere.

Lighting should be limited. Axolotls are sensitive to bright light and prefer gloomy environments. If you have live plants that require light, use modest aquarium lights and provide lots of shaded areas in the tank for the axolotl to hide. Timed illumination that mimics the natural day-night cycle can also be helpful.

Water quality parameters should be carefully monitored.

Axolotls are extremely sensitive to ammonia, nitrites, and nitrates. Investing in a decent water testing kit helps you monitor these levels and keep them within acceptable limits frequently. Before adding your axolotl, the tank should be cycled appropriately to ensure that helpful bacteria have established themselves to break down waste products. Regular water changes, typically 20-30% each week, help to maintain water quality and reduce the accumulation of dangerous contaminants.

A secure lid is also required for an axolotl aquarium. Although axolotls are not known to jump frequently, they try to escape occasionally. A tight-fitting lid minimizes mishaps and keeps your pet safe within the tank.

Feeding tools such as tongs or dishes might be beneficial. Axolotls are carnivorous and eat worms, tiny fish, and specially prepared pellets. Feeding equipment can assist in keeping the procedure clean and prevent leftover food from contaminating the water.

Setting up an axolotl tank necessitates attention to detail and an awareness of their unique requirements. You can provide a comfortable and healthy habitat for your axolotl using the appropriate tank size, substrate, decorations, perfect water conditions, and regular maintenance. Watching your axolotl thrive in a well-set-up tank is a beautiful experience emphasizing the necessity of thorough preparation and continuing care.

SUBSTRATE AND DECORATIONS

Creating the ideal environment for an axolotl requires careful consideration of the substrate and decorations in their tank. These features not only enhance the aesthetics of your aquarium but also play an essential part in your aquatic pet's health and comfort.

Choosing the correct substrate is the first step toward creating an axolotl-friendly environment. Axolotls are bottom dwellers who spend most of their time wandering down the substrate. As a result, it is critical to choose a material that is both safe and comfortable for them. Fine sand is often regarded as the best option. It resembles their native habitat and is easy for them to maneuver on. Unlike gravel, fine sand minimizes the possibility of impaction, a severe ailment caused by an axolotl accidentally ingesting substrate particles while feeding. On the other hand, Gravel might constitute a serious health danger since axolotls may swallow it when foraging, causing obstructions in their digestive tracts.

Following the installation of the substrate, the tank's decorations must be considered. These should fulfil functional and aesthetic reasons, resulting in an engaging habitat meeting the axolotl's requirements. Hiding areas are incredibly crucial. Axolotls are nocturnal species that prefer dimly lit environments where they can hide during the day. Providing children with numerous hiding places reduces stress and fosters a sense of security. This can be accomplished with various aquarium-safe decorations, including caves, tunnels, and P.V.C. pipes. These things not only provide shelter but also provide an aesthetic appeal to the tank.

Live plants can be an excellent addition to an axolotl's tank, providing benefits beyond décor. They oxygenate the water and can help maintain water quality by absorbing nitrates. Plants also help to create a more natural environment for your axolotl by offering cover and adding to its overall health. However, selecting resilient plants that can grow in the colder water temperatures preferred by axolotls is critical. Plants like java fern, anubias, and water wisteria are suitable. If live plants are not viable, high-quality artificial plants can be utilized with the same appearance.

When organizing the tank, try strategically adding decorations to create a diverse landscape. This method can provide your axolotl with a stimulating environment promoting natural activities like exploring and hunting. Smooth, rounded decorations are preferred over sharp or rough things that could harm the axolotl's sensitive skin and gills.

The overall layout and flow of the tank are also important considerations. Axolotls require open rooms for swimming and feeding. Hence, it is critical to alternate richly decorated portions with open ones. This gives them space to walk around and engage in natural behaviors. Arranging decorations to create mild slopes and walkways can make the tank more appealing and passable for your pet.

In addition to valuable goods, adding a few aesthetic

elements might improve your enjoyment of the tank. Decorative materials such as driftwood, rocks, and themed ornaments can add visual appeal while maintaining your axolotl's safety and comfort. Verify that any additional items are aquarium-safe and free of hazardous chemicals or sharp edges.

The substrate and decorations must be maintained regularly for the tank to function correctly. Gently agitate sand substrates regularly to avoid compaction and encourage healthy bacterial colonies. Decorations and plants, whether living or fake, should be cleaned regularly to remove algae and dirt. This not only keeps the tank clean but also avoids accumulating hazardous elements that could degrade water quality.

Creating an adequately constructed tank with the appropriate substrate and decorations improves your axolotl's physical and mental well-being. A simple aquarium can be transformed into a vivid, active ecosystem by adding fine sand, plenty of hiding spaces, and natural components such as plants. This meticulous attention to detail ensures that your axolotl flourishes, satisfying you to know you've provided a safe and exciting environment for your intriguing aquatic buddy.

Understanding the value of each piece and how it adds to the tank's overall environment allows you to enjoy building a beautiful and practical space that meets your axolotl's specific demands. Watching them explore their surroundings, interact with the decorations, and live contentedly in their carefully maintained home is a pleasant experience emphasizing tank design and maintenance.

FILTRATION AND AERATION SYSTEMS

Establishing an effective filtration and aeration system for your axolotl tank is critical to keeping a healthy environment. Axolotls are especially sensitive to water quality thus, clean water is vital for their health. Understanding the importance of filtration and aeration can help ensure that your axolotl lives in a stable and safe environment.

Filtration systems are essential for eliminating trash and preserving the chemical equilibrium of water. There are three forms of filtration: mechanical, biological, and chemical. Each has a distinct purpose in keeping the water pure.

Mechanical filtration is the first line of defense, removing material like uneaten food, excrement, and other particle matter. This is accomplished using a filter medium, such as sponges or floss, to catch these particles physically. Regular maintenance, such as cleaning or replacing the filter material, is required to avoid clogging and for proper operation.

Biological filtration is equally significant because it uses good microbes to break down hazardous pollutants. These bacteria colonize the tank's filter medium and other surfaces, converting poisonous ammonia from axolotl waste to nitrites, then converted to less damaging nitrates. This mechanism, known as the nitrogen cycle, is critical to maintaining a healthy aquatic environment. It is essential to allow these beneficial bacteria to establish, especially when starting a fresh tank. This process, known as cycling, might take many weeks but is necessary for establishing a stable environment.

Chemical filtration uses activated carbon or other chemical media to remove pollutants and poisons from water. While not usually necessary, chemical filtration can assist in eliminating any remaining contaminants, odors, or discolorations, resulting in crystal-clear water.

The appropriate filter for your axolotl tank is determined by the tank's size and your unique requirements. Canister filters, internal filters, and sponge filters are typical options. Canister filters outside the tank are solid and capable of handling more significant amounts of water, making them perfect for larger tanks. They have superior mechanical, biological, and chemical filtering properties. Internal filters are buried in the tank and are ideal for smaller installations, offering adequate filtration without taking up too much space. Sponge filters, noted for their soft water flow, are suitable for axolotls. They provide mechanical and biological filtration while ensuring the water flow does not cause stress to the axolotls.

Aeration is another crucial aspect of a healthy tank. Axolotls require oxygenated water, which aeration systems help to maintain. Air stones, diffusers, and air pumps are standard methods for increasing oxygen levels. These devices generate bubbles that improve gas exchange at the water's surface, allowing more oxygen into the tank. This is especially critical if your tank has a lid, as poor air exchange can lower oxygen levels.

Aeration also promotes water circulation, which can help distribute heat and nutrients evenly throughout the tank. Axolotls enjoy calm waters, so make sure the water flow isn't too forceful. Adjustable air pumps or strategically placed air stones can assist in controlling the flow and keep it gentle.

Regular monitoring and maintenance of your filtration and aeration systems is required. Checking the filter media, cleaning the components, and replacing parts as needed will ensure the system runs smoothly. Regularly testing water parameters such as ammonia, nitrite, nitrate, and pH levels ensures that filtration is successful and the tank environment is stable.

Maintaining a balanced and clean environment with adequate filtration and aeration improves your axolotl's general health. It inhibits the accumulation of dangerous pollutants and promotes a stable ecosystem. This maintains your axolotl healthy and enhances their quality of life by allowing them to engage in natural behaviors and thrive in their aquatic environment.

Setting up these systems may appear complicated initially, but they require reasonable maintenance once done. A well-maintained filtration and aeration system provides a pristine and healthy environment for your axolotl. Observing your axolotl in a clean, well-oxygenated tank is rewarding because you know you've created an environment that promotes their health and longevity. With the proper care and attention, your axolotl can live a long, healthy life in a stable, clean aquatic habitat.

LIGHTING AND TEMPERATURE CONTROL

Creating the right home for your axolotl requires attentive lighting and temperature control. Both of these features are critical in maintaining the health and comfort of these unusual amphibians. Axolotls flourish in specific environments, and understanding their demands is essential to providing the best care possible.

Lighting is one of the more manageable components, yet it is still quite crucial. Axolotls are nocturnal species that inhabit lakes and canals' dark, muddy waters. They enjoy low-light environments and might grow stressed when exposed to intense lights. They should avoid using bright, direct lights when setting up their aquarium. Instead, use muted lighting that replicates their natural habitat. L.E.D. lights are a fantastic alternative because they are energy-efficient and generate less heat, which can aid in maintaining the desired temperature.

Lighting is required for photosynthesis if you have live plants in your axolotl aquarium. However, you can still use low-intensity lights to give adequate illumination for plant development without overpowering your axolotl. Consider utilizing a timer to control the lighting cycle, replicating a natural day-night rhythm with approximately 10-12 hours of light per day. This uniformity promotes a stable environment and decreases stress for your axolotl.

The tank's positioning also affects light exposure. Keep the tank out of direct sunlight, warming the water and promoting algae growth. A corner of the room with indirect light is frequently the most excellent option for keeping the tank in a consistent, controlled light condition.

Temperature regulation, on the other hand, necessitates more active intervention. Axolotls are native to the chilly, steady waters of high-altitude lakes in Mexico, where they flourish at temperatures ranging from 60 to 68 degrees Fahrenheit. Maintaining this temperature range is critical, as variations can create stress and health problems.

Many homes, particularly those in warmer areas, need help to achieve and maintain the proper water temperature. An aquarium chiller is an excellent way to keep the water cool. These gadgets are intended to reduce the temperature of the water while keeping it steady. While chillers are an expenditure, they are frequently required to provide an appropriate atmosphere for axolotls.

Another option for regulating temperature is to use a fan to chill the water surface. Evaporative cooling from the fan can reduce tank temperature by a few degrees. This method is most effective with room air conditioning to keep the ambient temperature cooler. To ensure the temperature remains within the appropriate range, monitor it regularly with a dependable aquarium thermometer.

During the colder months, if your home temperature gets too low, you may need to utilize an aquarium heater. However, this is less prevalent because most residences keep the temperature above the minimum for axolotls. If you use a heater, be sure it has an accurate thermostat to avoid overheating the tank.

Maintaining water temperature stability is critical. Rapid temperature changes can cause thermal shock, which is hazardous to axolotls. Consistency is essential, and even minor variations should be eliminated. Regularly checking the temperature and making minor adjustments as needed will aid in maintaining a steady atmosphere.

Proper lighting and exact temperature control create a comfortable and stress-free environment for your axolotl. These factors affect your pet's physical health, behavior, and well-being. Axolotls in appropriate settings are likelier to exhibit natural behaviors, remain active, and have more enormous appetites.

The attention to detail in lighting and temperature control

demonstrates the effort needed to care for these intriguing creatures. Every component, from selecting the appropriate light intensity to investing in temperature control technology, helps create an environment as close as feasible to their natural home.

The payoff for this meticulous treatment is a healthy, thriving axolotl that will have a long and happy life in your care. Watching your axolotl explore its tank, engage in natural activities, and thrive in a well-maintained environment demonstrates the value of providing the proper conditions. Proper lighting and temperature control will help keep your axolotl healthy and active in your aquarium.

Chapter 3

SETTING UP THE PERFECT HABITAT

CREATING AN IDEAL ENVIRONMENT

WATER QUALITY: PARAMETERS AND TESTING

To create an optimum environment for an axolotl, ensure the water is of the highest quality. These tiny amphibians are susceptible to water conditions; thus, maintaining the proper parameters is critical to their health and lifespan.

The path to flawless water quality begins with understanding the essential parameters: ammonia, nitrites, nitrates, pH, and temperature. Each component substantially impacts the aquatic habitat's overall health and, thus, the axolotl's well-being.

Ammonia is the first parameter that needs to be closely monitored. It results from axolotl waste, which includes uneaten food and excrement. In a well-balanced aquarium, helpful bacteria convert ammonia to nitrites and nitrates. However, high ammonia levels can be dangerous. High ammonia levels can induce stress, gill damage, and even mortality in axolotls. As a result, ammonia levels must remain as low as feasible. Regular testing with an ammonia test kit ensures that any spikes are noticed early and dealt with quickly.

Nitrites, the second important parameter, are an intermediate product in the nitrogen cycle. Nitrites, while less poisonous than ammonia, can cause severe injury to axolotls by impairing their ability to transfer oxygen in the blood. Keeping nitrite levels at zero is optimal. This is accomplished with a wholly cycled tank where helpful bacteria convert nitrites to nitrates effectively.

Nitrates, the ultimate result of the nitrogen cycle, are less toxic but still require management. High nitrate levels might cause stress and long-term health problems. Regular water changes are the primary approach for lowering nitrate levels. Ideally, nitrate levels should be kept at 20 ppm. Consistent testing keeps track of nitrate levels and ensures that water changes are made as needed.

PH is another important factor in water quality. Axolotls prefer pH levels of 6.5 to 8.0. This spectrum is somewhat acidic and slightly alkaline. Maintaining a constant pH within this range keeps axolotls comfortable and lowers the risk of pH shock. Drastic pH changes can cause severe stress. Hence, it is critical to use a trustworthy pH test kit regularly. To prevent the axolotl from being shocked, make any necessary modifications gradually.

Temperature control is also essential. Axolotls flourish in cooler water, with an optimal temperature range of 60 to 68 degrees Fahrenheit. Warm water can cause stress, lower oxygen levels, and an increased risk of infection. In contrast, too-cold water can slow their metabolism. Regular monitoring with a dependable aquarium thermometer ensures the temperature stays within acceptable limits. If the indoor temperature is too high, an aquarium chiller may be required to keep the water at the optimum temperature.

Regular water testing is required to maintain these parameters. A thorough water test kit, including tests for ammonia, nitrites, nitrates, and pH, is a must-have for every axolotl keeper. Weekly water testing helps to spot any problems before they become serious. Keeping a record of these test results can also help you track changes and detect patterns that could signal a problem.

In addition to frequent testing, water changes are essential for maintaining water quality. Partial water changes, around 20-30% of the tank volume weekly, aid in removing excess nitrates and other waste materials. When changing water, use dechlorinated water because chlorine and chloramines in tap water can kill axolotls. To neutralize these toxins, use aquarium-specific water conditioners.

Creating an optimal environment also entails ensuring that the tank has been properly cycled before introducing an axolotl. Cycling a tank entails cultivating a constant population of beneficial bacteria capable of efficiently processing ammonia and nitrites. This procedure may take several weeks, but it is necessary for a healthy and balanced tank. Using a test kit to check ammonia, nitrite, and nitrate levels during this time can assist in deciding when the tank is ready for its new tenant.

Maintaining ideal water quality necessitates continuous monitoring, testing, and attentive maintenance. Understanding and maintaining ammonia, nitrite, and nitrate, pH, and temperature levels will help provide a stable and healthy environment for your axolotl. This effort assures their health and well-being and lets them thrive by displaying their distinct behaviors and bright personalities in a well-kept aquatic habitat.

CYCLING THE TANK: A STEP-BY-STEP GUIDE

Cycling the tank is a crucial step that all axolotl owners must learn and carry out before introducing their new pet. This technique creates a balanced ecology within the tank, allowing beneficial bacteria to adequately handle waste products while providing a safe and healthy habitat for the axolotl. Here's a step-by-step approach to cycling your tank and preparing it to host aquatic life.

The first step in cycling your tank is to set up the aquarium.

To remove dust or residue:

Begin by carefully cleaning the tank and equipment with dechlorinated water.

Avoid using soap and chemicals, as they might leave dangerous residues.

Once the tank is clean, place the substrate on the bottom, which should be fine sand for axolotls.

Arrange decorations, plants, and hiding places to provide a comfortable and fascinating environment for your axolotl.

Next, fill the tank with dechlorinated water. Tap water usually contains chlorine or chloramines, which are toxic to axolotls and the good bacteria you're attempting to grow. Use a water conditioner to purify the water and neutralize these pollutants. Monitor the temperature while you fill the tank to ensure it stays between 60 and 68 degrees Fahrenheit, which is perfect for axolotls.

After filling the tank, it's time to install the filtration system. A good filter is vital for sustaining water quality because it provides mechanical, biological, and occasionally chemical filtering. Axolotls enjoy tranquil waters, so choose a filter that is appropriate for the size of your tank and provides a gentle flow. Once operational, the filter will aid in establishing and maintaining the nitrogen cycle.

Now, the cycle process begins. The idea is to grow a colony of helpful bacteria that converts hazardous ammonia, created by fish waste and uneaten food, into nitrites and less toxic nitrates. There are two main strategies for accomplishing this: fishless and fish-in cycling.

Fishless cycling is preferable since it protects axolotls and other fish from potentially hazardous ammonia and nitrite levels. To begin fishless cycling, you will require a supply of ammonia. This can be pure ammonia added to the aquarium or organic food like fish or shrimp. If using pure ammonia, add a modest quantity daily to keep the ammonia concentration between 3-5 ppm (parts per million). If you're using organic material, put it in a mesh bag to keep it from breaking apart and making a mess in the tank.

Test the water daily with an aquarium test kit to check ammonia, nitrite, and nitrate levels. Initially, ammonia levels will rise, followed by an increase in nitrites as the bacteria convert the ammonia. Nitrites will rise and then drop as they are transformed into nitrates. This procedure can take several weeks, but it is critical to be patient and allow the bacteria colonies to develop fully.

During this time, continue to add ammonia and test the water. Your tank has cycled when the ammonia and nitrite levels drop to zero within 24 hours of adding ammonia, and the nitrate levels rise. At this stage, a significant water change (approximately 50%) should be made to lower the nitrate concentration to safe levels, ideally below 20 ppm.

If you opt for the fish-in cycling technique, which is not suggested because of the stress it can give to live fish, you will need to introduce a few robust fish to produce ammonia organically. To keep ammonia and nitrite levels low enough to protect the fish, test the water regularly and change it out frequently. This approach necessitates close observation and more frequent maintenance.

Regardless of the approach, once the tank has been cycled, it is critical to sustain the bacterial colonies by supplying a consistent source of ammonia, which will come from your axolotl's waste when it is introduced. Regular testing and water changes are still necessary to maintain the stability of the tank's ecosystem.

Introducing your axolotl to a wholly cycled tank can reduce stress and ensure a healthy start in your new home. The time and work put into correctly cycling your tank will pay off by creating a safe and stable environment in which your axolotl can fly. By following these instructions, you ensure your tank is ready to support life, resulting in a dynamic and balanced aquatic home for your intriguing pet.

- PLANTS AND HIDES: CREATING A STIMULATING HABITAT

Creating a fascinating habitat for your axolotl requires careful design and attention to their natural activities and needs. Plants and hides are crucial components of a well-designed tank, providing functional benefits while improving overall beauty.

Begin by choosing the appropriate plants for your axolotl's house. Live plants are a fantastic alternative because they add a natural feel to the tank and improve the water quality. They absorb nitrates, oxygenate the water, and provide more hiding places for your axolotl. Java fern, anubias, and water wisteria are excellent plant selections for axolotl tanks. These plants are sturdy and can tolerate the colder water temperatures that axolotls love. They also perform well in low-light environments, making them ideal for tanks where bright illumination is avoided to keep the axolotl comfortable.

Java fern is popular since it can be connected to rocks and driftwood, allowing for unique positioning in the tank. Anubias, with its large leaves, provides excellent cover and is easily attached to decorations. Water wisteria spreads swiftly and produces dense foliage, which can create a lush environment. These plants decorate the tank and provide a more natural and enriching environment for your axolotl.

If caring for live plants seems too tricky, artificial plants are another choice. High-quality fake plants can seem like natural plants and offer similar benefits regarding hiding places and visual appeal. When selecting artificial plants, ensure they are constructed of safe, non-toxic materials and have no sharp edges that could injure your axolotl.

Hides are another essential feature of the habitat. Axolotls are naturally shy creatures who prefer solitary regions where they can retreat and feel protected. Providing a diversity of hides reduces stress and promotes natural behaviors. There are numerous hides, including caves, tunnels, and mainly built axolotl hides. These can be fashioned from aquarium-safe materials, including resin, ceramic, or P.V.C.

Axolotls enjoy hiding in caves because they provide a dark and secluded environment to rest. Commercially available caves come in various forms and sizes, or you may make your own out of materials such as clay pots or coconut shells. Tunnels made of P.V.C. pipes or other smooth materials can also be good hiding places and pathways for your axolotl to explore.

When arranging hides and plants in the tank, try to create a balanced pattern that includes open swimming regions and thickly grown sections. This design allows your axolotl to wander freely while providing access to numerous hiding locations. Placing plants and hides strategically throughout the tank promotes exploration and enables your axolotl to engage in natural activities such as foraging and hunting.

In addition to plants and hides, consider adding other components that will improve the environment. Smooth rocks, driftwood, and colorful ornaments can all contribute to a visually appealing and exciting environment. To avoid mishaps, make sure any additional decorations do not have sharp edges and are firmly affixed.

It is also crucial to keep the tank clean and the plants healthy. Regular trimming of live plants promotes their health and prevents them from overloading the tank. Cleaning decorations and hides after water changes contributes to a clean environment, lowering the danger of algae growth and other potential problems.

The mix of plants and hides adds to the tank's appeal and considerably increases your axolotl's life. A stimulating environment encourages physical activity, lowers stress, and improves general well-being. Watching your axolotl explore, hide, and interact with their surroundings is a gratifying experience emphasizing the tank's value.

Developing a stimulating environment is a constant

endeavor. As you observe your axolotl and learn more about their preferences, you can make further changes and improvements to improve their surroundings. This meticulous care keeps your axolotl healthy, active, and interested, allowing you to appreciate their distinct charm and characteristics. A well-planned tank with the correct plants and hides strikes the ideal balance of beauty and usefulness, providing a safe and enriching environment for your axolotl.

MAINTAINING WATER QUALITY

REGULAR WATER CHANGE

Maintaining water quality is one of the most critical components of axolotl care, and regular water changes are vital to this process. The health and well-being of these rare amphibians depend on keeping the water clean and clear of dangerous contaminants.

Axolotls are sensitive organisms whose water environment

must be carefully regulated to prevent waste accumulation. Ammonia, nitrites, and nitrates are the most common waste products that can be collected in an axolotl tank. These compounds can be harmful, causing stress, disease, and even death if not appropriately managed. Regular water changes are the most efficient strategy to control these levels and maintain a healthy tank environment.

The method of making a water change begins with identifying your axolotl's individual needs and tank size. A weekly water change of approximately 20-30% of the tank's volume is advised. This frequency helps keep the water parameters consistent and minimizes the concentration of dangerous elements without inflicting undue stress on your axolotl.

To begin, gather all necessary tools, including a siphon or gravel vacuum, a bucket, and a water conditioner. The siphon removes water and debris from the tank, while the bucket collects the filthy water. The water conditioner is necessary because tap water frequently contains chlorine or chloramines, which are toxic to axolotls. Treating the new water with a conditioner neutralizes the chemicals and makes it safe for your pet.

I am using the siphon to drain water from the tank to begin the water change. Place one end of the siphon in the tank and the other in the bucket. Start the siphon and gradually suck the substrate to remove debris and waste. This method lowers the concentration of dangerous compounds and keeps the tank clean and visually appealing.

While siphoning, keep an eye on the axolotl and any decorations or plants. Move the siphon gradually to avoid overstimulating your axolotl since rapid movements and noise can create stress. Continue siphoning until you've removed the necessary amount of water, usually around 20-30%.

After removing the unclean water, it is time to prepare the fresh water. Fill a clean bucket with tap water and add the water conditioner according to the manufacturer's recommendations. This step is critical for removing toxic compounds that could harm your axolotl. Also, ensure the new water is the same temperature as the tank water to avoid scaring your axolotl.

Slowly pour the conditioned water back into the tank.

Pouring it in too rapidly may disrupt the substrate and stress your axolotl. Pouring the water over a rock or ornament can help spread the flow and minimize damage. Continue adding water until the tank has returned to its original level.

After doing the water change:

Take a minute to examine your axolotl.

Make sure there are no indicators of stress or discomfort.

Look for typical actions like swimming and exploring the tank. If everything appears good, your water change is done.

Regular water changes are more than simply a maintenance activity; they are essential to providing a stable and healthy habitat for your axolotls. Consistency is essential. Keeping a timetable and making water changes regularly helps prevent the accumulation of dangerous substances and preserves overall water quality.

In addition to water changes, regular testing of water parameters is required. Check the levels of ammonia, nitrites, nitrates, and pH regularly with a reputable water test kit. These tests provide helpful information about the tank's water quality, allowing you to make more educated decisions about maintenance and care.

Regular water changes and water quality monitoring ensure that your axolotl can grow in a clean and stable environment. This regimen helps to prevent common health disorders and encourages your pet's long, healthy life. By devoting time and effort to maintaining water quality, you provide a safe and nurturing environment where your axolotl can thrive.

HANDLING COMMON WATER ISSUES (AMMONIA, NITRITE, NITRATE)

Maintaining a healthy habitat for an axolotl necessitates careful monitoring of water quality, with ammonia, nitrite, and nitrate levels being the most important metrics to monitor. Each of these compounds is essential to the nitrogen cycle, and imbalances can harm your axolotl's health. Understanding how to deal with typical water challenges is critical for ensuring a stable and safe environment.

Ammonia is the first chemical to be recognized in the nitrogen cycle. It is made from axolotl excrement, leftover food, and decomposing organic substances. Ammonia in high concentrations is exceedingly poisonous to axolotls, causing stress, gill tissue burns, and potentially deadly conditions. The key to managing ammonia levels is to do regular testing. Check the water for ammonia at least once a week and more frequently if you suspect a problem.

If ammonia levels exceed zero, rapid action is required. Perform a 20-30% water change to dilute the ammonia levels. Ensure the new water is treated with a conditioner to remove chlorine and chloramines. In addition, inspect the tank for uneaten food or waste and remove it as soon as possible. Overfeeding is a typical source of high ammonia levels, so alter feeding techniques as needed.

Introducing helpful bacteria can also assist in controlling ammonia levels. These bacteria convert ammonia to nitrites, a less toxic chemical. Add products containing live bacteria to the tank to speed up this natural process. Ensuring that your filter is efficient and provides enough biological filtration is critical. Over time, the bacteria will colonize filter media and other tank surfaces, helping to keep ammonia levels under control.

Nitrite, the following step in the nitrogen cycle, is also poisonous to axolotls. It interferes with their ability to absorb oxygen, resulting in brown blood disease. Regular nitrite testing is crucial, especially during the tank's first cycling. Any detectable level of nitrite suggests an issue that requires attention.

If nitrite levels are high, perform a partial water change to reduce the concentration. Adding one teaspoon of aquarium salt per gallon can help reduce nitrite toxicity to axolotls by preventing it from being absorbed via the gills. Before adding the salt to the tank, ensure it is completely dissolved. Continue to carefully monitor nitrite levels and make further water changes as needed.

Beneficial bacteria are essential in the nitrogen cycle's last stage, converting nitrites to nitrates. Ensuring that your biological filtration is effective will aid in nitrite control. Products that promote bacterial growth can be helpful during the initial tank setup or when nitrite levels rise. Over time, the bacteria will convert nitrites to nitrates, which are significantly less toxic.

Nitrates, while not as poisonous as ammonia or nitrites, can create long-term health problems if left to collect. High nitrate levels can cause stress in axolotls and contribute to poor water quality. The most efficient method of controlling nitrate levels is regular water changes. Aim to keep nitrate levels below 20 ppm by changing 20-30% of the tank water weekly.

Live plants can also help control nitrate levels. Plants take nitrates as part of their growing process, which lowers the content in the water. Including tough, low-light plants like java fern or anubias in your tank will help control nitrate. Consider adding nitrate-absorbing material to your filter to reduce levels further.

Regular maintenance and monitoring are essential for preventing and resolving common water concerns. Establishing a weekly testing schedule for ammonia, nitrite, and nitrate levels will help you avoid possible problems. A diary of test results can provide significant insights into the patterns and changes in your tank's water quality, helping you make more educated maintenance and care decisions.

Managing ammonia, nitrite, and nitrate levels requires regular testing, frequent water changes, and adequate biological filtration. Understanding the nitrogen cycle and taking proactive actions to preserve water quality can provide a stable and healthy environment for your axolotl. This diligence guarantees your pet flourishes and is free of poor water quality, stress and health hazards. Providing a clean and balanced environment lets your axolotl show natural behaviors and enjoy a long, healthy life under your care.

Chapter 4

FEEDING YOUR AXOLOTL

NUTRITIONAL NEEDS

NATURAL DIET IN THE WILD

To understand axolotl nutritional demands, one must investigate their natural diet in the wild. These intriguing amphibians, which are native to Mexico's Xochimilco and Chalco lake complexes, have evolved to exist on a specific set of food sources that support their distinct lifestyle. By closely duplicating this diet in captivity, we can ensure that axolotls get the nutrition they need to stay healthy and active.

In the wild, axolotls are predatory predators. They hunt mainly at night, using their excellent senses to find prey.

Their diet consists of microscopic aquatic organisms that supply a high concentration of protein and other nutrients. In the wild, axolotls primarily eat insects, worms, tiny fish, and crustaceans.

Worms, especially earthworms and bloodworms, are a staple of the axolotl's diet. These give a high-protein, low-fat supper that axolotls may easily digest. Earthworms are very nutritious, including vital amino acids and lipids that benefit the axolotl's overall health. Bloodworms, the larvae of midge flies, are another excellent source of nutrients. They are high in protein and iron, sustaining the axolotl's energy levels and overall vigor.

Insects play an essential role in the axolotl's natural diet. They eat various aquatic insects and their larvae, including mosquito larvae and small beetles. These insects supply protein and essential vitamins and minerals for growth and development. Chitin in insect exoskeletons, for example, can assist digestion while providing a tiny quantity of dietary fiber.

Small fish are also an essential food source for wild axolotls. They feed on small fish that live in the same lakes and canals, such as minnows and guppies. These fish provide a healthy combination of protein, lipids, and other nutrients. Axolotls get omega-3 and omega-6 fatty acids from eating small fish, essential for healthy skin and gills.

The cuisine also includes crustaceans such as shrimp and little crayfish. These contain protein and a significant amount of calcium and other minerals required for bone and cartilage formation and maintenance. While challenging to digest in large quantities, the hard shells of crustaceans are ingested in moderation, and help wear down the axolotl's teeth, which grow continuously throughout their lifetimes.

Understanding these food habits is critical to delivering enough nourishment in captivity. To simulate their natural diet, a range of foods should be provided. Earthworms, bloodworms, and black worms provide great staple foods that may be served daily. These are widely accessible at pet stores or can be grown at home with little effort. Small feeder fish can be added sometimes, but they must be free of diseases and parasites that could harm the axolotl.

In addition to these natural food sources, specially-made

axolotl pellets can augment the diet. These pellets are intended to give balanced nourishment, containing all of the critical vitamins and minerals required for healthy growth. However, they should not be used to replace live or frozen feeds entirely, as axolotls benefit from the variety and stimulation provided by hunting and eating live prey.

Feeding axolotls in captivity should mirror their natural hunting behavior. Offering food items individually and watching the axolotl's reaction guarantees that they eat correctly and enjoy their meals. Removing any uneaten food after feeding is critical to preserve water quality and prevent the accumulation of dangerous waste products.

Regular feeding regimens are necessary. Young axolotls, who are still increasing, should be fed every day. Adults can be fed every other day to ensure they are getting enough nutrition without overeating, which can lead to obesity and other health problems. Monitoring their weight and modifying portions accordingly helps them stay in good health.

Understanding and reproducing axolotls' natural diet allows us to ensure that they get the nourishment they need to thrive in captivity. This technique benefits their physical health and improves their quality of life by enabling them to engage in natural behaviors while maintaining their colorful, vibrant personalities. One of the most critical components of axolotl care is to provide them with a varied and balanced food, which contributes to their overall health and lifespan.

THE BEST FOODS FOR CAPTIVE AXOLOTLS

Feeding captive axolotls the most excellent foods is critical to their health and happiness. While they have distinct nutritional demands based on their natural diet, giving a range of high-quality foods ensures they receive all the nutrients they require to flourish.

Earthworms are a favorite and highly nutritious diet for axolotls. Earthworms are a natural choice because they are incredibly healthy, easy to digest, and widely available. They are high in protein required to grow and maintain body processes. Additionally, earthworms supply essential lipids and amino acids. Earthworms can be purchased from bait shops, pet stores, or homes on worm farms. Feeding earthworms to axolotls is as simple as putting them into the tank, where they will be eagerly hunted and consumed.

Bloodworms are another great choice. These little, crimson midge fly larvae are protein-rich and ideal for immature axolotls. Bloodworms can be obtained frozen or freeze-dried, which is the better option because it maintains more nutrients. Thaw frozen bloodworms in tank water before feeding them to your axolotl. This keeps the water temperature from decreasing dramatically, which could stress the axolotl. Bloodworms can be fed with tweezers or placed in the tank, where the axolotl will find them.

Black worms, like bloodworms, are a highly nutritious alternative. They are slim, watery worms that axolotls cannot resist. Black worms are a high-protein food that is especially beneficial to young axolotls. These can be obtained live or freeze-dried; however, live black worms promote more natural hunting behaviors. They can be placed immediately in the tank, where they will wiggle enticingly, catching the axolotl's eye.

Axolotl pellets, designed explicitly for nutritional

requirements, are a practical and dependable feeding source. These pellets are intended to float or sink slowly, replicating the natural movement of prey in the water. High-quality axolotl pellets are nutritionally balanced, with the proper ratio of proteins, lipids, and vitamins. They are a fantastic supplement to a diversified diet, ensuring your axolotl gets all the nutrients it needs. When picking pellets, seek those with few fillers and a high protein content.

Small fish, such as guppies or minnows, can be added to the diet. These should be fed sparingly and only if they are disease and parasite-free. Feeder fish can be an excellent source of protein and vital fatty acids. They also provide enrichment by boosting the axolotl's innate hunting impulse. When introducing feeder fish, ensure they are quarantined for some time to prevent any health issues from entering the tank.

Shrimp, particularly ghost shrimp or little prawns, are another tremendous culinary option. These crustaceans are high in protein and have a unique texture and flavor, providing diversity to the diet. Live shrimp can be placed immediately in the tank, where they swim around, enticing the axolotl to hunt. Frozen shrimp are also OK but must be thawed before feeding.

In addition to these essential food sources, it is vital to include other delights to create a well-rounded diet occasionally. Daphnia, sometimes known as water fleas, are microscopic crustaceans that make good food for juvenile axolotls. They have a high protein content and can be fed live and frozen. Tubifex worms, albeit less prevalent, are another treat that can be given on occasion.

Feed your axolotl according to their size and age. Juveniles require more regular feedings, usually once daily, although adults can be fed every other day. Overfeeding should be avoided because it might lead to obesity and poor water quality. Observing your axolotl's bodily state and modifying portions accordingly aids in maintaining optimal health.

A varied diet serves the nutritional demands of axolotls and keeps them interested and active. Watching them search and engage with various foods may be an enjoyable experience for the pet and the owner. Cleaning the tank regularly and discarding uneaten food helps maintain water quality and minimizes potential health problems.

A variety and balanced diet can help your axolotl thrive. Every meal provides an opportunity to watch their fascinating behavior and ensure their health and vitality. This careful attention to their dietary demands promotes their overall well-being, helping them to live a long and meaningful life under your care.

FEEDING PRACTICES

HOW OFTEN TO FEED

Feeding your axolotl correctly is critical to their health and growth. One of the most crucial components of feeding is deciding how frequently to feed them. Understanding feeding frequency benefits their health, prevents overfeeding, and ensures they get enough nutrition.

Like many other animals, Axolotls have varying food requirements throughout their lives. Young axolotls, less than six months old, are rapidly growing and require more frequent feeding. At this point, it is ideal to feed them every day. Offering them food once a day guarantees that they acquire enough nutrition to support their rapid growth and development. Young axolotls have a faster metabolism thus daily feeding helps meet their energy needs.

When feeding juvenile axolotls, offering a variety of high-protein options is crucial. Bloodworms, black worms, and little chunks of earthworm are all wonderful alternatives, whether live or frozen. These feeds should be coarsely minced to accommodate the juvenile axolotls' small size. Feeding them in the evening or at night can replicate their everyday feeding habits, as axolotls are nocturnal hunters in the wild.

As axolotls mature, their growth rate slows, and feeding frequency must be regulated. Adult axolotls above six months benefit from eating every other day. This timetable helps avoid overfeeding, leading to obesity and water quality issues. Adult axolotls can be fed a more comprehensive range of meals, including larger earthworm pieces, axolotl pellets, and occasional treats such as small fish or shrimp.

Feeding adult axolotls every other day ensures they get enough nutrition without storing too much food in their tank, which can rot and pollute the water. Maintaining a clean tank is critical for the health of your axolotl, and regulating the amount of food brought into the tank is an integral part of that process.

Watch your axolotl's behavior and overall health when creating a feeding schedule. Each axolotl is unique, and some may need to be fed more frequently than others. Axolotls that appear underweight or lose weight may need to be fed more regularly, whilst those who gain weight may benefit from fewer frequent meals.

Look at your axolotl's body shape to see if they get enough food. A healthy axolotl's body is spherical and about the same width as its head. If the axolotl appears overly thin, with a notable drop in body mass, consider increasing feeding frequency or portion size. If the axolotl appears bloated or has excessive fat deposits, limit the amount and frequency of feeding.

Portion control should also be part of your feeding routines.

Offering modest amounts of food and monitoring how much your axolotl consumes within a few minutes will help prevent overfeeding. To maintain the quality of the water, any uneaten food should be removed from the tank immediately. This approach keeps the tank clean and prevents hazardous bacteria and fungi from growing on leftover food.

In addition to regular feeding, adult axolotls can benefit from intermittent fasting days. Skipping one meal every week can imitate the erratic dietary habits that axolotls may encounter in the wild. This method reduces obesity and encourages the axolotl to stay active and attentive.

It's also worth noting that axolotls' feeding habits vary depending on the season and surroundings. During the colder months, their metabolism may slow, and they will eat less regularly. Changing the meal schedule to match their lower activity levels is critical.

Understanding how often to feed your axolotl and modifying their food based on age, size, and activity level can help them stay healthy and vibrant. Regularly examining their behavior and bodily condition helps to fine-tune feeding procedures, ensuring they receive adequate nutrients without overfeeding.

Feeding your axolotl adequately benefits their general health by boosting growth, reducing disease, and improving their quality of life. Whether you're caring for a young, rapidly growing axolotl or an adult who needs less frequent feedings, consistency and attentiveness are essential for successful axolotl care. This strategy guarantees that your axolotl flourishes, allowing you to watch them grow and prosper in a well-managed, balanced habitat.

PORTION SIZE AND FEEDING TECHNIQUES

Feeding your axolotl in the proper portion sizes and employing appropriate feeding strategies are critical to guaranteeing their health and well-being. Understanding how much to feed and how to present food will help prevent overfeeding, preserve water quality, and promote your axolotl's natural habits.

Portion sizes differ according to the age and size of your

axolotl. Young axolotls under six months old must be fed little portions regularly. Juvenile axolotls proliferate and have high metabolic demands. A good rule of thumb is to provide enough food that they can finish in a few minutes. This usually equates to a small pinch of bloodworms or a tiny bit of earthworm. It is preferable to begin with smaller amounts and grow as needed rather than risking overfeeding.

As axolotls age, their growth rate decreases, and their food requirements shift. Adult axolotls beyond six months should be fed more significant portions less regularly. A portion size for an adult axolotl can be around the size of its head. This could refer to a giant earthworm, many pellets, or a small fish. The goal is to supply enough food to suit their nutritional demands while avoiding surplus that will decay and pollute the water.

Feeding strategies are also crucial for ensuring that your axolotl gets enough food. One efficient way is to hand-feed with tweezers or feeding tongs. This method gives you control over the portion size and ensures that the meal is devoured, reducing wastage. Hold the food near the axolotl's mouth, and they will quickly eat it. Hand feeding also allows you to monitor your axolotl's health and hunger closely.

Another option is to drop the food directly into the tank. This approach is practical for live and frozen meals, such as bloodworms or black worms. To foster natural foraging behavior:

Scatter the food throughout different places.

If you feed pellets, keep an eye on the quantity because uneaten pellets deteriorate quickly and harm water quality.

Place a few pellets at a time, ensuring they're swallowed before adding more.

Using a food dish can also help keep the tank clean. Place the food in a shallow dish near the bottom of the tank. This limits the food to a single area, making it easy to remove any uneaten bits and preventing them from spreading around the tank. Feeding bowls benefit axolotls, who are less active or have trouble finding food.

Regardless of the feeding method, it is critical to develop a routine. Consistency allows your axolotl to anticipate feeding times and relieves stress. Feed young axolotls daily and adults on alternate days. Please keep track of their physical health and alter portion quantities as necessary. An axolotl's body should be rounded but not too chubby. If your axolotl is gaining too much weight, try reducing the portion size or feeding frequency.

Offering a range of foods to promote a well-balanced diet is also beneficial. Alternate between earthworms, bloodworms, black worms, and top-quality pellets. This variety covers their nutritional requirements and keeps them interested in their meals. Avoid feeding fatty or processed foods, as they might cause health problems.

In addition to regular feedings, adult axolotls can benefit from intermittent fasting. Skipping a meal once a week helps to prevent overfeeding and reflects natural eating habits in which food availability varies. Fasting days relieve the axolotl's digestive tract and can improve general health.

Maintaining water quality is critical when adjusting portion sizes and feeding methods. Uneaten food can quickly degrade water quality, causing ammonia surges and discomfort for your axolotl. After feeding, discard any uneaten food within a few hours. Regular water changes and monitoring of water parameters contribute to a clean and healthy atmosphere.

Feeding your axolotl correctly, with appropriate portion sizes and practices, promotes their health and longevity. Observing their reactions to different foods and changing their approach ensures they get enough nourishment without overfeeding. This meticulous care promotes a healthy, lively axolotl, improving their quality of life and bringing joy to their caretakers.

LIVE FOOD VS. PREPARED FOOD

Choosing between live and prepared food for your axolotl requires a grasp of the advantages and disadvantages of each option. Both types offer benefits, and striking the perfect mix can considerably improve your axolotl's health and well-being.

Live food has various benefits, making it an essential part of an axolotl's diet. Axolotls are energetic hunters in their natural environment, feeding on various tiny animals.

Offering live food in captivity allows them to engage in these natural behaviors, providing physical and cerebral stimulation. Worms, particularly earthworms and black worms, are excellent live food sources. They are high in protein and easy to digest, and their twisting motion piques the axolotl's interest, encouraging a healthy feeding response.

In addition to worms, live insects like crickets and small fish like guppies can be used. These meals provide variety in the diet and ensure that the axolotl receives a broad spectrum of nutrients. Live food also appeals more to picky eaters, promoting a healthy appetite.

However, feeding live food comes with its own set of complications. One of the main worries is the possibility of introducing parasites and viruses into the tank. Getting live food from reliable providers that guarantee that their products are free of dangerous diseases is critical. Quarantining live food before feeding it to your axolotl can assist in reducing these risks. Another disadvantage is handling live food, which needs more work than prepared food. Live food must be kept alive and healthy, which may require additional time and resources.

Prepared meals provide convenience and uniformity. High-quality axolotl pellets are designed to provide balanced nutrition, ensuring your pet gets all the essential vitamins and minerals. These pellets are intended to sink gently, simulating the natural movement of prey and encouraging feeding. Prepared foods are simple to store, have a long shelf life, and prevent the possibility of introducing parasites or infections.

Frozen meals like bloodworms and brine shrimp are also excellent prepared options. These feeds are often flash-frozen to retain nutritional value and are widely accessible at pet retailers. They provide a nutritional profile comparable to live meals without the trouble of keeping them alive. Thawing frozen food before feeding is straightforward and keeps it at a safe temperature for your axolotl.

While prepared foods are handy, relying primarily on them may have drawbacks. Axolotls used to eating live prey may be less interested in prepared dishes, resulting in a diminished appetite. Furthermore, certain prepared foods may contain fillers or lower-quality components with a different nutritional value than real food. To provide the healthiest diet for your axolotl, buy high-quality brands and verify ingredient lists.

Choosing the correct balance of live and prepared foods can provide the best of both worlds. Combining live food, such as earthworms or black worms, with high-quality cooked foods can result in a well-rounded diet. This method ensures that axolotls get various nutrients and can engage in natural hunting activities while benefiting from the convenience and nutritional value of prepared foods.

Feeding practices should be tailored to the axolotl's age, size, and health status. Juvenile axolotls require frequent feedings and may benefit from a higher proportion of live food to help them increase. Adult axolotls, who need less regular feedings, can thrive on a balanced diet of live and prepared items, with occasional treats for variety.

Regardless of the meal, monitoring your axolotl's reaction and modifying it accordingly is critical. Observing their eating habits, growth, and overall health can provide helpful information about whether their diet meets their requirements. Regularly cleaning the tank and discarding uneaten food helps to preserve water quality, which is critical for the axolotl's health.

Axolotls consume both live and prepared foods. Live food promotes natural behaviors and is highly nutritional, whereas prepared food gives convenience and balanced nutrients. Combining both sorts allows you to produce a diversified and enriching diet that promotes your axolotl's health and well-being, resulting in a long and vibrant existence.

Chapter 5

HEALTH AND WELLNESS

RECOGNIZING COMMON HEALTH ISSUES

SIGNS OF A HEALTHY AXOLOTL

Caring for an axolotl is more than simply giving food and shelter; it is critical to recognize and understand the indicators of a healthy axolotl to ensure its survival. Healthy axolotls exhibit certain behaviors and physical traits that show they are flourishing. Knowing these indications can help you spot potential health issues and take proper action.

A healthy axolotl is often lively and responsive. Although axolotls frequently sleep at the bottom of their tanks, they should also show signs of activity, especially during feeding times. A healthy axolotl will immediately become attentive and interested when food is introduced into the tank. They may swim toward the food or engage in hunting behaviors such as patiently stalking their target before capturing it. This could indicate a health issue if your axolotl is constantly lethargic or uninterested in feeding.

Clear, vivid gills are another vital sign of a healthy axolotl. The gills should be feathery and vividly colored, usually pink or red, suggesting proper oxygenation. Gill filaments should be complete and spread out, not clumped or shriveled. Any changes in the look of the gills, such as a loss of color or considerable shrinkage, can indicate stress or poor water quality and must be handled immediately.

A healthy axolotl's skin should be smooth, with no blemishes, discoloration, or growths. While minor abrasions may occur, mainly when the axolotl explores sharp or rough surfaces, their skin should be clear and intact. Axolotls have a protective mucus layer on their skin that helps them avoid infections. This layer should be visible but not too thick or flaking. Detecting strange patches, redness, or fungal development could indicate anything is amiss.

A healthy axolotl has a well-proportioned body that seems rounded and slightly chubby. Their limbs should be firm and properly developed, with no symptoms of swelling or deformity. The axolotl's body should be proportional to its head, with no significant weight loss or severe thinness. An axolotl that is too slim may be underfed or have an internal condition, whereas an axolotl that is too fat may be overfed or have metabolic difficulties.

Axolotl's eyes should be clear, with no cloudiness or puffiness. While some axolotl morphs have naturally diverse eye hues, any rapid eye clarity or appearance change should be constantly examined. Swollen or clouded eyes may suggest an infection or poor water quality, necessitating quick treatment.

Regular feeding is another indication of a healthy axolotl. They should actively accept food and be interested in their meals. If an axolotl refuses food for an extended period, it could be due to stress, disease, or poor water quality. Monitoring their feeding habits regularly is critical to ensure they are eating correctly.

Behavioral indicators are also helpful in determining the

health of your axolotl. They should be able to swim freely and effortlessly. If you detect inconsistent swimming, uncontrollable floating, or difficulties sustaining buoyancy, there may be a health problem. Axolotls occasionally float to the top to breathe, which can indicate concerns such as gas accumulation, illnesses, or poor water quality.

Healthy axolotls should produce normal faeces. Observing their excrement can reveal information about their digestive health. Typical axolotl excrement is firm and black. Observing diarrhea, a lack of waste, or any other abnormalities could indicate digestive or nutritional concerns. Providing a well-balanced meal with adequate quantities can aid in maintaining normal digestive function.

The water quality heavily influences the health of your axolotl in the tank. Regular testing for ammonia, nitrites, nitrates, and pH promotes a healthy atmosphere. Ammonia and nitrite levels should be nil, with nitrates at 20 ppm. The pH should range between 6.5 and 8.0. Clean, well-filtered water at the proper temperature (60 to 68 degrees Fahrenheit) promotes general health while lowering the risk of illnesses and stress.

Interaction with tank mates, if any, can indicate health. Axolotls should cohabit harmoniously with others of their size. Aggression, excessive hiding, or evidence of harm from tankmates may suggest stress or incompatibility. Providing ample hiding spots and monitoring interactions contributes to a peaceful tank habitat.

Regular observation and maintenance are essential for detecting and resolving health issues early. Setting aside time each day to observe your axolotl's behavior and physical condition might help you identify potential problems before they worsen. Keeping a journal of their food habits, activity levels, and changes in appearance might help them track their health over time.

A healthy axolotl is active, with bright and vibrant gills, smooth skin, a well-proportioned physique, clear eyes, and a strong appetite. Monitoring their behavior, physical condition, and water quality ensures their health. Understanding and identifying these signals can help you give the most fantastic care for your axolotl, allowing them to enjoy a long, healthy, and happy life.

COMMON DISEASES AND SYMPTOMS

With their distinct charm and fascinating nature, Axolotls demand close attention to their health. Like any other pet, they can develop various ailments, and detecting the symptoms early on can make a big difference in treatment and recovery. Understanding common diseases and associated symptoms will aid in keeping your axolotl healthy and happy.

Fungal infections are one of the most common health issues among axolotls. These typically manifest as white, cotton-like growths on the skin, gills, or fins. The fungus generally develops when the axolotl's immune system is impaired, frequently caused by stress or poor water quality. If you spot these white areas, respond soon. Begin by increasing water quality, doing a partial water change, and ensuring ideal temperature and pH levels. Antifungal medications created exclusively for aquatic animals can be put in the tank, but read the instructions carefully to prevent causing further stress to your axolotl.

Bacterial infections are another prevalent problem, and they can cause red, inflammatory spots on the skin, gills, or mouth. These diseases are frequently caused by contaminated water or infected wounds. If you see these signs, place the affected axolotl in a quarantine tank with pristine water conditions. Antibiotics supplied by a veterinarian with experience with amphibians can help treat the infection. They identify and fix the underlying problem, such as sharp objects in the tank or elevated ammonia or nitrite levels.

Axolotls can be infected with external and internal parasites as well. External parasites, such as anchor worms, can attach to the skin or gills, causing irritation and distress. Internal parasites, such as protozoa, can produce symptoms such as weight loss, fatigue, and an enlarged abdomen. Parasite treatment necessitates a thorough diagnosis, which a veterinary professional frequently performs. External parasites can occasionally be removed manually or treated with anti-parasitic drugs. Internal parasites may require a different sort of medicine delivered through food.

Bloating is one of the most dangerous disorders that axolotls can develop. Bloat causes the axolotl's body to enlarge, making it appear abnormally puffy. Several factors, including bacterial infections, kidney failure, or intestinal obstructions, could cause this. If your axolotl appears swollen, you should immediately consult a veterinarian. They can identify the underlying reason and prescribe appropriate treatment, such as antibiotics, dietary adjustments, or other medical measures.

Axolotls are also prone to stress difficulties. Poor water quality, inappropriate temperature, and bullying by tank mates are all potential sources of stress. Stress-related symptoms include loss of appetite, frequent hiding, and changes in color or gill condition. Improving tank conditions, such as guaranteeing clean water, giving appropriate hiding places, and maintaining a constant temperature, can help reduce stress. If tank mates create stress, consider separating them or providing a larger tank to prevent territorial issues.

Poor nutrition can contribute to various health problems, including metabolic bone disease caused by a lack of calcium and vitamin D3. Symptoms may include soft or malformed bones, trouble swimming, and drowsiness. This condition can be avoided by feeding your axolotl a well-balanced meal rich in calcium and ensuring they have access to adequate lighting or supplements.

Another typical issue is impaction, which occurs when the axolotl consumes substrate or other things that clog the digestive tract. Symptoms include loss of appetite, lethargy, and a bloated abdomen. To avoid impaction, use fine sand or bare-bottom tanks instead of gravel. If impaction develops, see a veterinarian. They may prescribe soaking the axolotl in shallow water to let the obstruction pass, as well as other therapies.

Maintaining high water quality is critical for avoiding several of these ailments. Regularly test the water for ammonia, nitrites, nitrates, and pH. Keep the tank clean with regular water changes, and keep the temperature between 60 and 68 degrees Fahrenheit. Avoid abrupt changes in water conditions, which can stress your axolotl and make it more prone to sickness.

Being concerned about your axolotl's health and identifying the symptoms of common ailments early can significantly impact their well-being. Regular inspection, keeping excellent water quality, providing a balanced diet, and getting veterinarian guidance are all essential for safeguarding your axolotl's health and well-being in its aquatic environment. Understanding and resolving potential health issues early can allow you to enjoy the companionship of a bright and healthy axolotl for many years.

PREVENTATIVE CARE

QUARANTINE PROCEDURES

Ensuring your axolotl's health and wellbeing entails treating illnesses as they occur and implementing preventative actions. Implementing quarantine protocols is one of the most effective preventive care strategies available. Quarantine acts as a protective barrier, shielding your existing tank residents from potential diseases and parasites introduced by newcomers.

When you bring home a new axolotl or any other aquatic

species, the anticipation of adding it to your existing tank might be overpowering. However, patience and caution are necessary. Quarantining new arrivals before introducing them to the main tank is critical. This process usually lasts 30 days, giving enough time to observe and guarantee the immigrant is healthy and free of contagious infections.

Setting up a quarantine tank is simple. Begin with a separate, smaller tank with a filter, heater (if required), and hiding places. This tank should be kept in a quiet, stress-free atmosphere so the new axolotl can settle in comfortably. Ensure the water conditions in the quarantine tank match those in the main tank to reduce stress during the shift.

When the quarantine tank is completed, carefully place the new axolotl using a clean net or container to prevent cross-contamination. The quarantine tank's water quality must be regularly monitored. Perform regular water testing to ensure ammonia, nitrite, nitrate, and pH levels are within safe limits. Frequent water changes are required to maintain optimal conditions, as a smaller volume of water might get polluted faster.

During quarantine, keep an eye out for any symptoms of disease or suffering in the new axolotl. Common symptoms to look for include lethargy, loss of appetite, strange swimming patterns, changes in skin or gill appearance, and apparent parasites. If any of these symptoms arise, it is critical to address them quickly. A veterinarian who specializes in amphibians can advise on proper treatments and care.

Feeding the isolated axolotl should correspond to the food plan in the main tank. Provide a range of high-quality meals, including earthworms, bloodworms, and axolotl pellets. This provides sufficient feeding and allows you to assess the new axolotl's appetite and overall health. A healthy axolotl will usually exhibit interest in food and eat quickly.

In addition to monitoring health, the quarantine time allows the new axolotl to adjust to its new surroundings without the stress of interacting with established tank mates. This acclimatization period can be beneficial in minimizing stress-related illnesses and providing a smoother transition when it comes time to introduce the new addition to the main tank.

After the quarantine period is completed and the new axolotl shows no signs of disease, it is time to prepare for its introduction into the main tank. Before moving, perform a last health check and a partial water change in the quarantine and main tanks to ensure the highest possible water quality. Over several hours, small volumes of water from the main tank will be gradually introduced into the quarantine tank to adapt the new axolotl to the water in the main tank. This allows the axolotl to adjust to any minor variations in water chemistry.

When moving the axolotl to the main tank, use a clean net or container to prevent infection. Introduce the new axolotl cautiously and watch for any signs of hostility or discomfort from the existing tank dwellers. Providing a variety of hiding places and visual barriers can help prevent territorial disputes and allow the new axolotl to choose a comfortable home.

Quarantine precautions are required for new axolotls and anyone sick or injured who must be isolated for treatment. Having a quarantine tank on hand allows you to offer necessary care without jeopardizing the health of the other tank dwellers. It also enables close monitoring and more accurate therapy through medication, dietary changes, or environmental modifications.

Implementing comprehensive quarantine protocols can drastically limit the chance of bringing diseases and parasites into your axolotl tank. This proactive strategy promotes a healthy and stable habitat for all aquatic creatures, ensuring their wellbeing and longevity. Taking the effort to quarantine new arrivals indicates your dedication to proper pet management and the overall health of your aquatic community.

HOW TO TELL IF YOUR AXOLOTL IS SICK

Understanding whether your axolotl is sick is critical to their health and wellbeing. These fascinating amphibians require special care, and recognizing early signs of disease can make a big difference in their recovery. Here's a complete guide to determining whether your axolotl is sick and what steps you should take to keep them healthy.

Axolotls are noted for their toughness and remarkable regenerating ability but are nevertheless vulnerable to various health problems. Regular observation is essential for identifying any changes in behavior or appearance that may suggest a problem. Start by creating a routine for checking on your axolotl. This practice should include tracking their activities, hunger, and physical state.

A behavior change is often one of the first signs that anything is amiss. Healthy axolotls are usually active, particularly during feeding times. If your axolotl appears lethargic or spends more time hiding than usual, this could indicate stress or disease. Lethargy can be induced by various circumstances, including poor water quality, low temperature, or underlying health conditions.

Appetite is another vital sign of health. Axolotls are typically avid eaters, so a sudden loss of interest in food is a warning sign. If your axolotl refuses to eat for several days, you should examine deeper. Stress and environmental changes, as well as more severe problems such as infections or digestive disorders, can all cause a loss of appetite.

The physical appearance of an axolotl provides various

signals regarding its health. Healthy axolotls have smooth, even skin and vivid gills. Any skin texture or color changes, such as the appearance of white patches, sores, or lesions, may suggest an infection or fungal problem. Similarly, gills should be fluffy and vibrant; if they appear pallid, faded, or decaying, they indicate something is wrong.

Floating is another phenomenon to be aware of. While axolotls occasionally float, frequent or continuous flotation can signal digestive issues or distress. If your axolotl struggles to stay immersed, it could be due to bloating or impaction. Ensuring that kids eat a nutritious meal and do not consume substrate can help prevent these problems.

Water quality is critical for an axolotl's well-being wellbeing. Poor water quality is a leading cause of stress and sickness. Regularly test the water for ammonia, nitrites, nitrates, and pH. Ammonia and nitrite levels should always be nil, while nitrates should be below 20 ppm. Maintaining a steady pH between 6.5 and 8.0 is also critical. If the water parameters are incorrect, make a partial water change and inspect your filtration system to ensure it functions correctly.

Temperature is another essential consideration. Axolotls thrive in chilly water, ideally between 60 and 68 degrees Fahrenheit. Temperatures exceeding 72 degrees increase stress and make people more prone to sickness. If you reside in a warm climate, consider utilizing an aquarium chiller to keep the temperature steady and calm. Rapid temperature fluctuations can also be detrimental, so make slow adjustments as needed.

Stress is a significant cause of sickness in axolotls. Tank mates, limited hiding spaces, and inappropriate handling can all contribute to stress. Axolotls like quiet environments with few interruptions. Ensure their tank has plenty of hiding spots, such as caverns or plants, where they can go if they feel threatened. If you have other pets or young children, place the tank in a calm location where the axolotl will not be continuously disturbed.

Recognizing the symptoms of particular ailments can also help you diagnose and treat your axolotl. Common health problems include fungal infections, bacterial infections, and parasites. Fungal infections frequently appear as white, cotton-like patches on the skin or gills. These infections are usually treated with salt baths or antifungal drugs. Bacterial infections can produce redness, swelling, or ulceration and are generally treated with antibiotics. Parasites may cause your axolotl to scratch at objects in the tank or exhibit unusual behavior. Identifying the individual parasite is critical for successful therapy, which frequently includes medication.

When you feel your axolotl is sick, consult a veterinarian specializing in amphibians. They can make an accurate diagnosis and provide appropriate therapies. Finding an amphibian-experienced veterinarian can be difficult, so it's best to investigate and create a relationship with one before an emergency occurs.

In addition to detecting disease symptoms, it is critical to be informed of legal issues surrounding axolotl ownership and treatment. Laws and regulations governing axolotls can differ significantly depending on your location. For example, certain states in the United States require permits to keep axolotls due to their status as an endangered species in the wild. To prevent potential legal complications and to support the conservation efforts of these beautiful species, make sure to follow local regulations.

Preventative care is the most effective way to keep your axolotl healthy. Clean the tank regularly, evaluate the water quality, and supply a well-balanced feed. Avoid overfeeding since leftover food can decay and pollute water quality. To prevent illness from spreading, quarantine new tank additions before bringing them into your main tank. Also, avoid using harsh chemicals or untreated tap water, which can be hazardous to axolotls.

Understanding disease symptoms and how to respond are essential components of ethical and responsible axolotl ownership. Regular monitoring, correct tank maintenance, and respect for legal standards will guarantee that your axolotl is healthy and thrives in your care. Being proactive and aware may create the most significant habitat for your axolotl, ensuring many years of companionship with this magnificent creature.

REGULAR HEALTH CHECKS

Keeping your axolotl healthy takes persistent care and

attention, with frequent health checks being an essential part of that routine. These exams help you detect any signs of illness, allowing for timely intervention and effective treatment. You can keep your axolotl healthy and happy by including health checks into your daily care regimen.

Begin by observing your axolotl's behavior and activity level. A healthy axolotl should alternate between rest periods and active swimming, particularly during feeding. They should respond promptly to food and engage in normal hunting activities. If your axolotl appears lethargic, uninterested in eating, or behaves abnormally, it may suggest a health issue that requires attention.

Scrutinize the skin and gills. The skin should be smooth and devoid of blemishes, discoloration, and growth. Axolotls with a healthy mucus layer have a better chance of avoiding illnesses. You should act immediately if you observe any anomalies, such as white spots (which could suggest fungal infections) or redness and irritation. The gills should be bright and fluffy, typically pink or red, indicating proper oxygenation. Gills that are pale, shriveled, or showing necrosis symptoms may indicate low water quality or disease.

Check the axolotl's eyes regularly. They should be clear, without cloudiness or puffiness. While certain axolotl morphs have naturally varying eye hues, unexpected changes in clarity or appearance may indicate a sickness or injury. Swollen or clouded eyes need prompt attention since they can compromise the axolotl's vision and overall health.

Evaluate your axolotl's physical condition. A well-nourished axolotl has a rounded body about the same width as its head. They should not appear too thin or puffy. Weight loss, a sunken appearance, or bloating may suggest food disorders, intestinal parasites, or other health problems. Regularly monitoring their weight and changing their diet helps them maintain good health.

Feeding patterns provide essential information about your axolotl's health. A healthy axolotl has a robust appetite and is eager to eat. Changes in eating patterns, such as refusing food or spitting it out, might be early indicators of disease. Ensure their diet is diverse and balanced, with live items such as earthworms, bloodworms, and high-quality pellets. Observe their eating habits and make any necessary adjustments to ensure they are getting enough nourishment.

Inspect the axolotl's legs and tail. These should be fully developed, with no evidence of swelling, deformity, or injury. Healthy limbs and tails are critical to their mobility and overall well-being. Any irregularities in these locations may suggest damage or disease. Axolotls have excellent regenerative ability, but any injuries must be monitored to promote appropriate healing and prevent infection.

Water quality is an integral part of regular health assessments. Regularly test the water for ammonia, nitrites, nitrates, and pH. Ammonia and nitrite levels should always be zero, as even trace amounts can be toxic. Nitrate levels should not exceed 20 ppm, and the pH should be between 6.5 and 8.0. Maintaining clean, well-filtered water at a consistent temperature (60 to 68 degrees Fahrenheit) is critical for avoiding stress and sickness.

Regular tank cleaning and water changes help to maintain optimal water quality. To avoid accumulating dangerous elements, immediately dispose of any uneaten food and garbage. Maintaining a clean and stable environment benefits your axolotl's health by lowering the chance of infections and stress-related disorders.

Interactions with tank mates, if any, should be noted. Axolotls should cohabit harmoniously with others of their size. Aggression, frequent hiding, or apparent injuries from tank mates suggest stress or incompatibility. Providing plenty of hiding locations and monitoring interactions might assist in establishing a peaceful tank environment.

Documenting your axolotl's health and behavior can be beneficial. Keep a notebook to record their eating patterns, weight fluctuations, and any changes in appearance or behavior. This record can provide valuable insights and aid in identifying patterns that may suggest health problems. It also helps veterinarians make more informed treatment decisions.

Regular health checkups can help keep your axolotl healthy and happy. Early discovery of potential health conditions allows for timely treatment, increasing the likelihood of complete recovery. This proactive approach to care reflects your dedication to your axolotl's well-being and provides them with the highest possible quality of life.

TREATING ILLNESSES

FIRST AID AND EMERGENCY CARE

Treating ailments in axolotls necessitates close observation and timely treatment. When you realize something is wrong with your aquatic pet, you should seek first aid and emergency care immediately. Understanding the fundamentals of immediate therapy will significantly improve your axolotl's recovery.

One day, you may notice your axolotl acting particularly lethargic, with reduced interest in food and possibly apparent signs of distress, such as inflamed skin or clamped gills. These symptoms can be concerning, but understanding how to react promptly will help stabilize your axolotl until you can seek more veterinarian care.

First, isolate the infected axolotl. A quarantine tank is required in such cases. This separate tank should be pre-filled with clean, dechlorinated water and kept at the same temperature as the main tank, between 60 and 68 degrees

Fahrenheit. This controlled atmosphere decreases stress and prevents infections from spreading to other tank dwellers.

Next, the water quality in both the primary and quarantine tanks will be assessed. Axolotls frequently experience distress due to poor water conditions. Check for ammonia, nitrites, nitrates, and pH levels. Ideal conditions include zero ammonia and nitrite levels, nitrate levels below 20 ppm, and a pH between 6.5 and 8.0. If the levels are incorrect, make quick water modifications to restore equilibrium. Clean, toxin-free water is essential for healing.

A salt bath can be an excellent first aid treatment for fungal infections, typically manifest as white, cotton-like growths on the skin or gills. To dissolve non-iodized salt, use one to two teaspoons per liter of dechlorinated water. Place your axolotl in this solution for about 10 minutes, keeping a close eye on it to ensure it does not become stressed. The salt kills the fungus while soothing the damaged regions. Repeat this therapy daily until you observe results, but stay within 10 minutes per session to avoid injuring your axolotl.

Bacterial infections, characterized by red, inflammatory regions, can be more challenging to treat. Keep the quarantine tank clean, as clean water is essential for healing. Antibiotics may be required, but you must first speak with a veterinarian who can prescribe the appropriate prescription and dose. Over-the-counter therapies can be dangerous without proper supervision. Therefore, professional counsel is essential.

Parasites, both internal and external, are another common problem. External parasites, such as anchor worms, can be removed manually using tweezers if visible and accessible. Make sure to do this lightly to avoid harming the skin. Internal parasites, which produce symptoms such as weight loss and bloating, may require medicine. Anti-parasitic therapies should be administered under the supervision of a veterinarian, who can offer the safest and most effective methods.

Axolotls occasionally experience impaction, which occurs when eaten substrate or other debris obstructs the digestive tube. Symptoms include a bloated abdomen and loss of appetite. If you suspect impaction, try immersing your axolotl in shallow, chilly water to calm its digestive system. Avoid feeding for a day or two to allow the impaction to heal spontaneously. If the situation does not improve, veterinarian treatment may be required.

When dealing with physical injuries like cuts or abrasions, keeping the affected area clean is critical. Infections are reduced by using clean water and living in a stress-free environment. For mild injuries, add a small amount of aquarium salt to the water (approximately one teaspoon per gallon) to aid recovery. More severe injuries should be checked by a veterinarian, who may prescribe topical antibiotics or other therapies.

Another crucial element of emergency care is to keep your axolotl hydrated and stress-free. Stress can exacerbate any medical condition, making healing more difficult. To reduce stress, keep the quarantine tank in a quiet, dimly lit environment. Monitor the water temperature regularly to guarantee stability, as changes can cause additional stress.

In some circumstances, you may need to provide drugs directly. This should always be done with veterinary supervision. Proper dose is critical for any antibiotic, antifungal, or anti-parasitic treatment. Both overdosing and under dosing can be detrimental. Follow the veterinarian's instructions carefully and finish the entire course of therapy, even if your axolotl appears to recover faster, to ensure the illness is completely removed.

It is critical to monitor the patient regularly during the recovery phase. Keep a watchful eye on the axolotl's behavior, eating, and overall health. Observe any changes, improvements, or worsening of symptoms. This information might help your veterinarian change medications as needed.

Furthermore, it is critical to prevent further health problems after your axolotl has recovered. Evaluate and improve the entire habitat, including water quality, nutrition, and tank setup. Reduce tensions and follow a regular cleaning and maintenance program. A well-balanced, nutrient-dense meal will also assist your axolotl stay healthy.

Many common axolotl ailments can be efficiently managed and treated by keeping a close eye on them, acting quickly, and using appropriate first aid. Building a relationship with an experienced veterinarian who understands amphibians is essential for providing the best care for your axolotl. With the proper technique, you can help your axolotl recover from illness and live a long, healthy life under your care.

10 THINGS YOU SHOULD NEVER DO TO AXOLOTL

When your axolotl is sick, it's critical to understand what acts to avoid to avoid exacerbating the situation. Here are ten things you should only do while your axolotl is well, along with explanations to help you provide the best care possible.

First and foremost, never ignore the symptoms of sickness. Axolotls are generally hardy creatures, although they can become unwell if the surroundings are not optimal. Symptoms such as tiredness, loss of appetite, odd floating, and skin lesions indicate a problem. Ignoring these warning signs can result in a worsened disease or even death. Promptly addressing changes in behavior or appearance is critical for timely treatment and recovery.

Second, refrain from making significant adjustments to the water in the tank. While water quality is vital, abruptly changing the water can startle your axolotls. An abrupt change in water parameters can stress the animal, which is already vulnerable. Instead, more minor, frequent water changes should be made to gradually enhance the water quality without generating extra stress.

Avoid using untreated tap water. Tap water frequently contains chlorine and chloramines, which are toxic to axolotls. Always use a water conditioner to treat tap water before putting it in the tank. Chlorinated water can be highly harmful, worsening any existing health issues in your axolotl.

Never change the tank temperature abruptly. Axolotls thrive in chilly water, preferably between 60 and 68 degrees Fahrenheit. Rapid temperature swings can stress them out and impair their immune system. Use an aquarium thermometer to check the water temperature and make any changes gradually to prevent shocking your axolotl.

Avoid handling your axolotl excessively. Handling can be stressful, especially when they are unwell. Stress can slow their recovery and lead to additional difficulties. If you must move your axolotl, like during a tank cleaning or to a quarantine tank, do it carefully and with minimal handling.

Avoid using powerful drugs without competent counsel. Over-the-counter medications may appear to be a quick fix, but when taken incorrectly, they can often cause more harm than good. Certain drugs can harm axolotls or upset the delicate balance of their tank environment. Before providing any therapies, always speak with an experienced veterinarian who has worked with amphibians.

Avoid overfeeding your axolotl. When an axolotl is unwell, it may have a smaller appetite. Uneaten food can quickly foul the tank, lowering the water quality and worsening health issues. To keep a clean environment, serve modest, manageable servings and remove any uneaten food immediately.

If your axolotl is unwell, never introduce new tankmates. Introducing new species into the tank can raise stress and the danger of infection. New tank mates may introduce viruses or parasites, jeopardizing your axolotl's health. Maintaining a sick axolotl in a calm, lonely area is recommended until it has fully recovered.

Avoid making significant changes to the tank arrangement. While you may be tempted to extensively clean and rearrange the tank, this can be very stressful for your axolotl. Keep the surroundings as stable and familiar as possible. Make just the alterations required to preserve cleanliness and water quality without disturbing the tank's layout.

Finally, pay attention to your legal and ethical responsibilities. Depending on your location, there may be unique restrictions governing the treatment and maintenance of axolotls, mainly if they are a threatened species in the wild. Ensure you implement treatments or interventions according to local wildlife laws and regulations. Consulting a veterinarian helps you provide the finest treatment possible and assures that you adhere to all legal requirements.

Understanding what not to do when your axolotl is ill is just as vital as knowing what steps to take. Avoiding these frequent mistakes will help establish a more supportive environment promoting healing and rehabilitation. To guarantee the best possible results for your axolotl, always prioritize gentle, consistent care and seek professional guidance as needed. Your axolotl can heal and grow with careful and intelligent care, bringing joy and curiosity.

WHEN TO SEE A VETERINARIAN

Caring for an axolotl entails knowing when to visit a veterinarian. While regular observation and at-home care can help with many ailments, some necessitate professional medical attention to ensure your pet's health and well-being. Knowing when to seek veterinarian assistance can significantly improve your axolotl's recovery and overall quality of life.

Consider observing that your axolotl isn't acting like itself. It may be less active, not eating as much, or exhibiting symptoms of physical pain. These changes, however subtle, can be early signs of a health problem. If your axolotl's behavior changes dramatically and does not improve within a day or two, it is necessary to visit a veterinarian.

One of the most common reasons to seek veterinarian guidance is a prolonged lack of appetite. Axolotls are typically avid eaters, so a sudden lack of appetite can be concerning. If your axolotl refuses to feed for more than a few days, it could be suffering from a severe health problem that necessitates professional diagnosis and treatment. Prolonged anorexia can cause serious health problems, such as malnutrition and decreased immunity.

Visible physical symptoms like swelling, sores, or abnormal growths require quick veterinary intervention. Swelling around the abdomen may suggest impaction, infection, or internal parasites. Lesions or ulcers on the skin may indicate bacterial or fungal illness. Tumors and other dangerous illnesses are examples of abnormal growths. A veterinarian can undertake diagnostic procedures, such as imaging or biopsies, to diagnose the reason and propose the best treatment.

Respiratory difficulties are another major worry. If your axolotl is constantly gasping for oxygen at the surface or its gills appear inflamed and less vibrant than usual, it may be experiencing respiratory distress. This might be due to poor water quality, diseases, or environmental factors. A veterinarian can evaluate your axolotl's respiratory health, test the water parameters, and recommend changes or therapies to address the issue.

Behavioral changes, such as increased hostility, irregular swimming, or continuous hiding, can all indicate health problems. While axolotls have distinct personalities and sometimes act unexpectedly, regular behavioral changes should be noticed. These behaviors may suggest discomfort, stress, or neurological disorders. A veterinarian can help you uncover the underlying problem and provide ways to improve your axolotl's surroundings and health.

Another critical reason to seek veterinarian care is if you see any indicators of internal or external parasites. Internal parasites cause weight loss, lethargy, and bloating, whereas external parasites can appear as visible creatures that adhere to the skin or gills. Parasites can cause serious injury if left untreated, and a veterinarian can recommend suitable anti-parasitic medicines as well as advice on how to avoid further infections.

Water quality issues, despite frequent maintenance, may necessitate veterinary intervention. Suppose you constantly struggle to maintain acceptable water parameters and detect reoccurring health issues in your axolotl. In that case, a veterinarian can advise you on more advanced water treatment alternatives and help you build a more successful care routine.

Accidents and injuries are another area in which veterinary care is essential. Axolotls can harm themselves on tank decorations or while interacting with other tankmates. You should immediately consult a veterinarian if you see any injuries, bruises, or missing limbs. Axolotls have exceptional regenerative ability, but severe injuries necessitate medical attention to avoid infection and promote appropriate healing.

Even if your axolotl looks to be in good condition, monthly check-ups with an amphibian-experienced veterinarian will help. Preventative care, such as health examinations and advice on diet, water quality, and tank conditions, can assist in identifying any problems before they become serious.

Being aware of changes in your axolotl's behavior and appearance and understanding when to seek veterinarian treatment are critical components of responsible pet keeping. Loss of appetite, apparent physical symptoms, respiratory concerns, behavioral changes, parasite signs, recurrent water quality issues, accidents, and injuries are all evidence that professional medical attention is required. Establishing a relationship with a skilled veterinarian guarantees that you have a trusted source for your axolotl's health, allowing you to keep your pet healthy and happy for years to come.

Chapter 6

BEHAVIOR AND INTERACTION

UNDERSTANDING AXOLOTL BEHAVIOR

NORMAL VS. ABNORMAL BEHAVIOR

Understanding axolotl behavior is critical for giving optimal treatment. By distinguishing between typical and abnormal behavior, you can keep your axolotl healthy and happy in its surroundings.

Axolotls are fascinating creatures with various behaviors that are entertaining to watch. In their natural environment, they are primarily nocturnal, becoming more active at night. During the day, they often repose near the bottom of the tank, generally in a shady or secluded location. This resting behavior should not be misinterpreted as lethargy or disease; it is merely part of their natural routine.

One of the most distinguishing characteristics of a healthy axolotl is its feeding habits. Axolotls are opportunistic feeders and should respond quickly to food. When you feed food, such as worms or pellets, a healthy axolotl will make swift, deliberate movements to catch and swallow the prey. This willingness to eat reflects their general health and well-being. If your axolotl is not interested in food, it could indicate something is amiss.

Another common habit is their manner of investigating their surroundings. Axolotls employ sensitive skin and a lateral line system to detect changes in water currents and vibrations. You may notice your axolotl swimming slowly around the tank, occasionally pausing to inspect a new object or décor. This inquisitive conduct is critical to their mental stimulation and general satisfaction.

Axolotls also exhibit distinct behaviors associated with their unique breathing system. They have gills and lungs, allowing them to breathe water and air. An axolotl will periodically come to the surface for a gulp of air. This behavior is only cause for concern once it occurs frequently, which may suggest a problem with water quality or oxygen levels.

While these actions are typical of a healthy axolotl, it is also critical to spot indicators of atypical behavior that could suggest a health problem. One such indicator is chronic floating or an inability to remain immersed. Axolotls may occasionally float, but they should be able to return to the tank's bottom quickly. Persistent floating may suggest gas buildup, impaction, or buoyancy control concerns.

Erratic swimming and frenzied movements are also considered aberrant activities. These actions could indicate stress, discomfort, or irritation. Poor water quality, improper tank mates, or a lack of hiding spots could all be contributing factors. Ensuring the tank environment matches their requirements can help decrease stress and promote normal behavior.

Loss of appetite is another vital sign of probable health problems. If your axolotl repeatedly refuses food or displays little interest in eating, it could be due to stress, illness, or digestive issues. It is critical to observe their feeding habits and obtain veterinarian assistance if the problem persists.

Skin and gill appearance changes can indicate inappropriate behavior. Healthy gills are colorful and fluffy, usually in pink or red. If the gills are pale, shriveled, or have white areas, this could suggest poor water quality, infections, or stress. Similarly, the skin should be smooth and devoid of blemishes and discoloration. Any changes in skin condition should be handled immediately to avoid further issues.

Another sign of odd conduct is excessive concealment or a lack of mobility. While axolotls spend a significant amount of time resting, they should also exhibit moments of activity, particularly during feeding times. If your axolotl is continually hiding or has a noticeable decrease in activity, it could be due to disease or environmental stressors.

Observing interactions with tankmates is also essential. Axolotls are generally solitary organisms that should not engage in hostile behavior toward one another. However, detecting indications of fighting, such as bite marks or missing limbs, means that the tank habitat is too tiny or lacks hiding places. Providing appropriate space and resources can assist in alleviating these concerns.

In addition to behavioral observations, continuous monitoring of water quality is required. Poor water quality can cause a variety of health issues and odd behaviors. Regular monitoring for ammonia, nitrites, nitrates, and pH levels ensures the water remains within safe limits. Cleaning the tank and changing the water regularly can significantly impact your axolotl's health and behavior.

Creating a stimulating environment is another method for encouraging appropriate behavior. Providing a variety of hiding places, plants, and decorations promotes natural exploration while reducing stress. Keeping the tank adequately sized and not overcrowded helps your axolotl feel safe and comfortable.

Understanding the distinction between normal and pathological behavior in axolotls is critical for their care. A healthy axolotl exhibits regular feeding, exploration, occasional air gulps, and a balanced rest and activity schedule. In contrast, chronic floating, inconsistent swimming, loss of appetite, skin and gill appearance changes, extensive hiding, and hostility toward tank mates all indicate possible issues. Regularly monitoring your axolotl's behavior and keeping appropriate tank conditions can ensure they have a healthy and happy life under your care.

HANDLING AND INTERACTION TIPS

Handling and communicating with your axolotl necessitates a delicate approach and a thorough awareness of their needs. These unique amphibians are sensitive creatures, and proper interaction skills can help ensure their health and foster a bond with your pet.

Handling axolotls requires caution because of their sensitive skin and gills. To reduce stress and injury, direct handling should be restricted. The best way to relocate your axolotl is to use a soft, fine-mesh net. Gently sweep them up, ensure their gills and limbs are not trapped, and move them to their new place. Another efficient approach is to use a tiny container. Submerge the container in the tank, coax the axolotl inside, and lift it out. This lowers the danger of injury and stress.

If you need to handle your axolotl directly, ensure your hands are clean and clear of pollutants. Avoid using soap or hand sanitizer right before handling, as the residues can irritate their sensitive skin. Wet your hands with dechlorinated water to reduce skin damage and tension. Handle them with care, supporting their bodies without squeezing. Limit the time you hold them to reduce stress, and always return them to the water as soon and safely as possible.

Observing your axolotl's behavior can be a satisfying method to communicate without making direct touch. Spend time observing them explore their surroundings, eat, and rest. This not only helps you learn their typical activities but also allows you to see any disease symptoms or suffering early on. You can also improve their surroundings by changing decorations or introducing fresh plants and hiding places. This stimulates natural exploratory activity and keeps their surroundings interesting.

Feeding time provides another excellent opportunity for interaction. Axolotls usually are highly excited about food, and hand-feeding can help to establish trust. Use feeding tongs or tweezers to give them their preferred foods, such as earthworms or bloodworms. Hold the food close to their mouth and watch them gobble it up. This enriches their diet and allows you to monitor it and ensure they are eating correctly.

Axolotls require a calm and peaceful environment to function correctly. Loud noises, unexpected movements, and excessive handling can all lead to stress. Place their tank in a low-traffic area of your home so they feel safe. If you have other pets or small children, manage interactions to prevent the axolotl from becoming disturbed.

Establishing a schedule can also help your axolotl. Consistent feeding schedules and frequent tank maintenance contribute to a stable environment. Axolotls thrive on predictability, and routines assist in relieving stress and promoting good behavior. To keep the tank in good shape, check the water parameters regularly and clean it.

Enrichment is a crucial part of interaction. Adding living plants, driftwood, and caves creates a dynamic habitat.

Changing the arrangement of their tank might pique their interest and encourage exploring. Choose decorations that are safe for your axolotl and do not have sharp edges that could cause injury.

It's critical to understand when your axolotl requires space. Allow them to calm down if they withdraw to a hiding place or exhibit signs of stress, such as heated swimming or refusing food. Avoid handling children during these times and make sure their surroundings are conducive to relaxing.

Connecting with your axolotl necessitates a delicate approach and an emphasis on their requirements. When movement is required, utilize soft, fine-mesh nets or containers rather than direct handling. If you need to handle them, keep your hands clean and damp with dechlorinated water. Observing their behavior, hand-feeding, maintaining a peaceful environment, and offering enrichment are all effective ways to interact without generating stress. Establishing routines and detecting indications of discomfort can help keep your axolotl healthy and happy. Understanding and respecting your axolotl's demands will allow you to form a strong bond and enjoy the unique experience of caring for these fascinating creatures.

ENRICHMENT ACTIVITIES

TOYS AND ENRICHMENT IDEAS

Creating a stimulating habitat for your axolotl requires more than just primary care. Enrichment activities are vital for their mental and physical health. By introducing numerous toys and enrichment ideas, you encourage their natural habits, reduce stress, and keep them happy and healthy.

Axolotls are curious creatures who thrive in an environment encourages exploration and variety. Adding a variety of hiding locations is a simple yet efficient method to enrich their tank. Caves, tunnels, and P.V.C. pipes are great possibilities. These hiding places create a sense of protection and promote natural behaviors such as exploration and rest. You can buy these products from pet stores or make your own with safe, non-toxic materials.

Another excellent enrichment suggestion is to place natural plants in the tank. Axolotls enjoy chilly, low-light settings, so plants like java fern, anubias, and water wisteria thrive. Live plants provide various benefits, including improving water quality by absorbing nitrates, providing hiding places, and creating a more natural, visually appealing environment. As your axolotl weaves among the leaves and rests among the vegetation, you may enjoy watching them interact with the plants.

Introducing floating things can also bring interest to your axolotl's surroundings. Floating logs and lily pads provide surfaces to rest near the water's surface. These devices sway slightly with the water currents, encouraging mild movement and exploration. To avoid injuries, ensure that any floating items are safe and have no sharp edges.

Feeding time provides an excellent opportunity for enrichment. Instead of just putting food into the tank, use feeding tongs or tweezers to hand-feed your axolotls. This strategy stimulates your mind and helps you and your pet form stronger bonds. You may even make feeding puzzles by hiding food in small containers or beneath rocks. This encourages your axolotl to exercise its hunting instincts and makes eating more enjoyable.

Axolotls enjoy digging; therefore, giving a soft foundation, such as fine sand, can be beneficial. They frequently burrow or sift through the sand, mirroring their usual behavior in the environment. Ensure the sand is clean and clear of sharp debris that could injure your axolotl.

Adding decorative components like driftwood or smooth stones can further improve the tank habitat. These decorations offer a diverse environment that promotes exploration. Axolotls may be used to rub against, hide behind, or rest against. Check that all decorations are aquarium-safe and have no sharp edges.

You can also employ gentle water currents to create a dynamic setting. Axolotls prefer quiet waters, but a slow-moving current from a low-flow filter can provide subtle shifts to engage their senses. Adjusting the filter output or adding a small air stone will produce moderate bubbles that will spark your axolotl's interest without causing stress.

Mirrors can be used sparingly as an enrichment tool. Placing a small mirror outside the tank briefly can pique your axolotl's interest as they watch their reflection. However, this should only be done on occasion to avoid producing unnecessary stress or territorial behavior.

Changing the tank arrangement regularly can provide new stimuli while also preventing boredom. Every few months, rearrange plants, hides, and decorations to keep your axolotl's environment fresh and intriguing. Ensure that changes are implemented gradually to prevent generating tension.

Finally, observing your axolotl's response to enrichment activities is critical. Please observe their behavior and alter your activities based on their responses. If an enrichment item appears stressful or uninteresting, replace it with something else. The idea is to establish a balanced habitat to keep your axolotl interested and satisfied.

Hideouts, living plants, floating objects, interactive feeding methods, soft substrates, and a variety of decorations can all help to enrich your axolotl's environment. By incorporating these aspects, you may encourage your axolotl's natural habits, improve their well-being, and have fun watching them explore and thrive in a dynamic, engaging environment.

CREATING A STIMULATING ENVIRONMENT

Providing a fascinating environment for your axolotl entails more than simply addressing their basic needs. It is about creating an environment that stimulates natural activities, keeps them interested, and enhances their general health. Understanding what makes a habitat appealing to these intriguing critters allows you to improve their quality of life dramatically.

Could you start with the tank itself? A large tank allows your axolotl to explore and move freely. Aim for a tank that can accommodate at least 20 gallons of water for one axolotl, with extra space for each additional axolotl. The more space they have, the more opportunity for play and discovery.

Substrate is a significant consideration. Axolotls enjoy digging, so a soft, fine sand substrate can resemble their natural habitat. Avoid gravel, which can be consumed and induce impaction. Sand permits them to burrow and sift through it, carrying out natural foraging habits.

Hiding areas are essential for instilling a sense of security and stimulation. Use a variety of hides, including caves, hollow logs, and P.V.C. pipes. These not only offer places to retreat and rest, but they also promote exploration. Axolotls in the wild seek out recesses and shaded spots, so imitating this in their tank makes them feel more at home.

Natural and artificial plants can improve the environment for your axolotl. Live plants such as java fern, anubias, and water wisteria flourish in low light and cool water, just as axolotls. These plants provide hiding places, improve water quality by absorbing nitrates, and foster a more natural, engaging environment. If live plants are not an option, high-quality artificial plants can offer similar benefits without maintenance.

Adding a variety of ornaments might also stimulate your axolotl. Smooth rocks, driftwood, and aquatic ornaments provide diversity to the tank, giving your axolotl an array of textures and surfaces to explore. To avoid injuries, ensure that no decorations have sharp edges. These components improve the tank's cosmetic attractiveness and provide physical enrichment.

Lighting should be chosen appropriately. Axolotls like dim lighting because they are usually nocturnal, and their eyes are sensitive to intense light. Create a comfortable environment by using subdued lighting and shaded locations. If you have living plants that demand more light, ensure your axolotl has many shaded places to retreat.

Water quality and temperature are essential for a stimulating atmosphere. Regularly check and manage water parameters to ensure that ammonia and nitrite levels are zero, nitrates are less than 20 ppm, and pH is between 6.5 and 8.0. Keep the water's temperature between 60 and 68 degrees Fahrenheit. Clean, well-oxygenated water promotes your axolotl's health and activity level.

Feeding time presents an opportunity for enrichment. To stimulate hunting behavior, alter the sites where food is left. Feeding tongs can offer live or frozen prey, such as bloodworms or earthworms, and observe your axolotl's natural predatory instincts at work. You may also make feeding puzzles by hiding food in little containers or under decorations to encourage your axolotl to forage.

Interactive components, such as floating objects, can also give stimulation. Floating logs and lily pads allow your axolotl to examine and relax near the water's surface. The subtle movement of these pieces with water currents lends dynamism to their surroundings.

Changing the arrangement of the tank regularly helps keep things interesting. Every few months, rearrange decorations, plants, and hides to open new paths and regions for exploration. This keeps the surroundings fresh and exciting, boosting your axolotl's curiosity.

Observation is essential for ensuring your axolotl thrives in their habitat. Keep an eye on their conduct and adjust to their reactions. Consider adding new pieces or changing the arrangement if they appear anxious or disinterested. The goal is to create a balanced atmosphere that promotes physical and mental wellness.

In summary, providing a stimulating habitat for your axolotl requires a combination of adequate room, proper substrate, various hiding areas, plants, decorations, controlled lighting, and enriching feeding techniques. By carefully examining these factors, you can create an environment that will keep your axolotl active, healthy, and happy.

Chapter 7

BREEDING AXOLOTLS

BREEDING BASICS

UNDERSTANDING BREEDING BEHAVIOR

Breeding axolotls can be an exciting and rewarding experience, but it takes a thorough understanding of their breeding habits and demands. These fascinating organisms have particular patterns and behaviors that indicate they're ready to reproduce, and creating the appropriate conditions can assist in ensuring a successful breeding process.

The first stage in breeding axolotls is ensuring the pair is healthy. Both males and females should be mature, which usually means they are at least 12 months old. Mature axolotls are typically 7-8 inches long. Breeding can be physically demanding, so keeping both axolotls healthy is crucial. Feed them a well-balanced protein-rich diet that includes earthworms and high-quality pellets to ensure they are in peak condition before breeding.

Understanding the environmental variables that promote breeding is critical. Axolotls procreate due to seasonal changes, mainly lower water temperatures, and more rain. To recreate these conditions in captivity, gradually reduce the water temperature in the breeding tank to roughly 60 degrees Fahrenheit. This can be accomplished with an aquarium chiller or by relocating the tank to a colder environment. In addition, raise the water level slightly and keep the tank clean and well-oxygenated.

Once the environmental conditions have been established, the male and female axolotls will be added to the breeding tank. To create a pleasant and natural environment, offer a variety of hiding spaces and plants. Axolotls might be shy at their first meeting, so providing them with areas to retreat can minimize stress and encourage natural behavior.

Axolotls' breeding habits are highly peculiar. The guy usually starts the process by performing a wooing display.

This consists of actions and behaviors intended to capture the female's attention. The male will frequently swim around the female, nudging her with his nose and performing a "dance" by swaying his body and tail. These activities aim to arouse the female and indicate his eagerness to mate.

If the female is receptive, she will follow the male and engage in the wooing ritual. The male will next deposit spermatophores, or little packets of sperm, into the tank's substrate. This is typically done in a row or small cluster. The male will then lead the female over the spermatophores, prompting her to collect them with her cloaca. After the female has gathered the spermatophores, fertilization takes place internally.

After fertilization, the female will lay eggs within 24 to 48 hours. She will carefully deposit each egg on the tank's surface, like plants, ornaments, or tank walls. A single female can lay hundreds to thousands of eggs, depending on size and health. There should be lots of egg attachment surfaces to ensure that the eggs are evenly distributed and not crowded.

Once the eggs are laid, the adult axolotls must be removed from the breeding tank to avoid devouring the eggs. Return the adults to their usual tank, ensuring they are well-fed and tend to follow the breeding procedure.

Now, the emphasis is on caring for the eggs. Axolotl eggs are relatively robust but need stable water conditions to mature correctly. Maintain a water temperature of 68 degrees Fahrenheit and keep the tank clean and well-oxygenated. Avoid rapid changes in water parameters since they might hurt egg development.

Within two to three weeks, the eggs will start to hatch. The exact timing can vary depending on the water temperature and conditions. Newly born axolotls, known as larvae, are small and fragile. They will first feed on their yolk sacs, which supply vital nutrients throughout the first several days of life. Once the yolk sacs have been exhausted, it is critical to introduce suitable food sources.

Baby axolotls require live food that is small enough to eat. Infusoria, microscopic water microorganisms, make an excellent first food source. You can introduce larger live meals like baby brine shrimp and micro worms as the larvae develop. Feeding the larvae multiple times per day is critical to assist their rapid growth and development.

Maintaining water quality is crucial during this time. Regular water changes, careful monitoring of water parameters, and adequate filtration are all required to keep the larvae healthy. Overcrowding can occur; thus, the larvae may need to be separated into different tanks as they grow to ensure that each has adequate space and resources.

As the axolotls develop, their diet will increasingly include more significant foods. Chopped bloodworms and finely chopped earthworms can be introduced as the larvae mature into juvenile axolotls. To ensure optimal development, their growth and health must be continuously monitored during this stage.

Breeding axolotls involves time, attention to precision, and a thorough grasp of their unique demands and behaviors. You may successfully produce these fascinating organisms and contribute to their continued preservation and study by providing the necessary environmental conditions, closely monitoring the breeding process, and taking adequate care of the eggs and larvae. The entire process, from courtship to raising the young, is a pleasant experience for any axolotl aficionado.

PREPARING FOR BREEDING

Breeding axolotls necessitates a detailed grasp of their requirements and behaviors and meticulous preparation and setup. Ensuring that the environment and the axolotls are prepared for breeding can significantly improve the chances of success and the health of both the parents and their progeny.

The first step in preparation for breeding is to verify that your axolotls are healthy and fully developed. Axolotls are typically ready to procreate when they are at least 12 months old and have grown to approximately 7-8 inches long. Health is vital. Thus, both axolotls should be in peak condition, with no symptoms of illness or stress. Feed them a high-protein diet before breeding to ensure their strength and vitality. Earthworms, bloodworms, and high-quality pellets are excellent choices for providing all the required nutrition.

Creating the appropriate habitat is critical for successful breeding. Begin by creating a specialized breeding tank. This tank should be kept isolated from its typical habitat to create a controlled environment where you can precisely monitor the process. A tank size of at least 20 gallons is recommended, with enough room for both axolotls to move around comfortably. Ensure the tank has a sound filtering system to keep clean water, but the water flow should be modest to avoid stressing the axolotls.

Water quality is crucial to breeding success. The water temperature should be progressively reduced to roughly 60 degrees Fahrenheit, as lower temperatures can cause breeding behaviors. You can accomplish this by utilizing an aquarium chiller or locating the tank in a more relaxed area of your home. Maintaining excellent water quality, with ammonia and nitrite levels at zero and nitrates below 20 ppm, is also critical. Regular water testing and modifications are required to keep these parameters steady.

The breeding tank should include plenty of hiding nooks and vegetation. Axolotls require areas to retreat and feel safe, which reduces stress and promotes natural behavior.

Create a complicated and exciting habitat by combining live or fake plants, caves, and P.V.C. pipes. These features provide shelter and act as surfaces for the female to lay her eggs.

Once the breeding tank is ready, add the male and female axolotls. It's typically a good idea to keep them separated at first with a tank divider, allowing them to adjust to their new environment without coming into direct contact. Remove the partition after a day or two and observe how they interact. The male often initiates courtship actions such as pushing the female and performing a "dance" by swaying his body and tail. These activities suggest that the male is prepared to reproduce.

During this time, keep a close eye on the axolotls for symptoms of violence or distress. The tank environment must remain quiet and suitable for breeding. If the female is receptive, she will follow the male and finally collect the spermatophores he leaves on the tank floor. Fertilization occurs internally; after a day or two, the female begins depositing eggs.

The female will carefully lay her eggs on plants, ornaments, and tank walls. Depending on size and health, she can lay anywhere from a few hundred to thousands of eggs. Giving enough surfaces for her to lay her eggs comfortably is critical. Once the egg-laying procedure is complete, the adults should be removed from the breeding tank to prevent them from devouring the eggs. Please return them to their original tank, ensuring they are well-fed and cared for following the breeding procedure.

The focus now changes to caring for the eggs and preparing the larvae to hatch. Maintain a water temperature of roughly 68 degrees Fahrenheit and keep the tank well-oxygenated and clean. Avoid rapid water condition changes since these can harm the developing eggs. The eggs hatch in two to three weeks, depending on the temperature and environmental conditions.

Once the eggs hatch, tiny larvae will emerge. These infant axolotls are delicate and require special care. For the first few days, they will feed on their yolk sacs, giving them all the nutrients they need. Once the yolk sacs have been exhausted, it is critical to introduce suitable food sources. Infusoria, or tiny live aquatic microbes, are good starter meals. As the larvae develop, they gradually introduce larger live meals like baby brine shrimp and micro worms.

Maintaining proper water quality is even more critical at this point. To guarantee a healthy habitat for the larvae, change the water regularly and monitor the parameters. Overcrowding can be an issue as the larvae grow, so try splitting them into various tanks to give them enough space and limit competition for food.

As the larvae grow into juvenile axolotls, their food requirements shift. Begin by adding finely chopped bloodworms and earthworms, gradually increasing the size of the food as they grow. Consistent feeding and clean water are critical to their quick growth and development.

Throughout this period, keep a watchful eye on the health and conduct of parents and children. Regular observation will assist you in identifying any possible concerns early on, allowing for timely intervention. Documenting the mating process, from courtship behavior to larval development, can provide helpful information and help you better future breeding attempts.

Breeding axolotls can be a gratifying experience because it

allows you to see these fascinating creatures throughout their life. With careful planning, attention to detail, and a commitment to giving the best possible care, you can successfully breed axolotls and enjoy watching them grow from eggs to adults.

RAISING AXOLOTL LARVAE

CARING FOR EGGS AND HATCHLINGS

Raising axolotl larvae from eggs is an intriguing and gratifying activity that demands meticulous attention and perseverance. The procedure starts with caring for the eggs, ensuring they have the finest possible environment to develop and hatch.

Once the female axolotl has laid her eggs, providing a stable and safe environment is critical. To avoid swallowing the eggs, carefully remove the adult axolotls from the breeding tank. This can be accomplished with a soft, fine-mesh net or by gently persuading them into a container for transport. Once the adults have returned to their typical environment, concentrate on the eggs.

Axolotl eggs are frequently attached to plants, ornaments, or tank walls. They are relatively hardy but require consistent water conditions to grow correctly. The optimal temperature for egg development is approximately 68 degrees Fahrenheit. Maintaining this temperature is critical, as variations might harm the eggs. An aquarium heater or chiller can help adjust the temperature if necessary.

Water quality is another significant consideration. Ensure that the water is pure and well-oxygenated. Regularly test for ammonia, nitrites, nitrates, and pH. Ammonia and nitrite levels should be zero, nitrates below 20 ppm, and pH between 6.5 and 8.0. Partial water changes of 20-30% each week can help maintain optimal water quality. When changing the water, be gentle to prevent upsetting the eggs.

Within two to three weeks, the eggs will begin to hatch. The water temperature and the surrounding environment determine the exact timing. You'll see tiny larvae emerge from the eggs, each holding a yolk sac that provides vital nourishment for the first few days. These larvae are pretty delicate, so handle them with care.

Initially, the larvae will be primarily immobile, eating on their yolk sacs. During this time, keep a constant eye on the water quality to ensure it stays stable. After three to five days, when the yolk sacs have been consumed, the larvae will become more active and begin looking for food.

Feeding the larvae is a vital part of their care. Because of their small stature, they need live food appropriate for their size. Infusoria, microscopic water microorganisms, make an excellent first food source. These can be grown at home or bought from an aquarium supply store. As the larvae develop, they gradually introduce larger live meals like baby brine shrimp and micro worms. Feeding the larvae multiple times per day is critical to assist their rapid growth and development.

Providing plenty of food is critical, but so is preserving water quality. Overfeeding might result in excess food rotting in the tank, lowering water quality and harming the larvae. Remove any uneaten food after feeding to help keep the water clean. Regular water changes are necessary during this time but must be done gently to avoid stressing the larvae.

As the larvae mature, they exhibit distinguishing axolotl characteristics such as external gills and limbs. This growth phase is quick, and within a few weeks, they will resemble miniature copies of adult axolotls. During this time, it is critical to monitor their progress regularly. Ensure they're all growing at the same rate and watch for any symptoms of illness or deformity.

Overcrowding might become a problem as the larvae develop. Initially, you might have hundreds of larvae in a single tank, but as they grow, room becomes restricted. Overcrowding can cause food competition and increased waste generation, putting water quality at risk. To avoid this, consider dividing the larvae into many tanks as they mature. This gives each larva enough room to grow and decreases the chance of disease transmission.

Monitoring the larvae's health is critical. Look for symptoms of stress or disease, such as fatigue, a low appetite, or unusual swimming behavior. If you see any problems, act quickly. This could include altering the water conditions, modifying the nutrition, or, in some situations, consulting with an amphibian-experienced veterinarian.

As the larvae grow and develop, their nutrition must be modified. Introduce finely chopped bloodworms and earthworms as they grow larger and can take more food. A diversified diet ensures kids acquire all the nutrients required for optimal growth.

When the larvae reach a length of around 1-2 inches, they are termed juveniles and can be cared for similarly to adult axolotls. At this point, they should be given a well-balanced diet of live items and high-quality pellets. Regularly check their growth and health, and replace the water regularly to keep the environment clean.

Raising axolotl larvae involves perseverance, devotion, and meticulous attention to detail. To ensure the most outstanding results, each stage of the process, from egg laying to hatching and juvenile development, requires specific care and conditions. By providing a stable habitat, sufficient diet, and regular monitoring, you can successfully raise axolotl larvae into healthy, thriving adults, helping to preserve and enjoy these unique and fascinating creatures.

LARVAL DEVELOPMENT AND CARE

Observing the development of axolotl larvae is a fascinating experience that demands close attention and good care. Each stage of its development, from hatching to juvenile, has distinct requirements and milestones. Understanding these phases allows you to provide the most excellent care possible, promoting their development into healthy adults.

When the axolotl eggs hatch, the larvae are tiny and vulnerable, with yolk sacs that provide immediate sustenance. During the first several days, these yolk sacs keep the larvae alive, allowing them to grow modestly without needing external sustenance. It is critical to maintain clean water conditions throughout this time. The water temperature should be kept consistent, ideally between 68 and 72 degrees Fahrenheit and water parameters should be regularly checked to guarantee optimal circumstances. Ammonia and nitrite levels should be nil, with nitrates around 20 ppm.

Once the yolk sacs have been consumed, the larvae become more active and demand more food sources. They are so tiny that they can only eat minuscule living food. Infusoria, a little water bacteria, is ideal for their early diet. These can be grown at home or bought from an aquarium supplier. Small, regular feedings throughout the day assist in meeting their nutritional requirements and promote rapid growth.

As the larvae develop, they gradually introduce somewhat larger food items. Baby brine shrimp and micro worms are great options. These small animals move around in the water, stimulating the larvae's innate hunting abilities. Feeding them live food at this stage is critical because it promotes active foraging and helps the larvae acquire predatory skills.

Maintaining water quality is crucial at this period of tremendous growth. Overfeeding can result in extra food decaying in the tank, drastically worsening water quality. To keep the environment clean and waste-free, perform regular water changes of 10-20% every few days. Use a gently siphon to remove trash while avoiding upsetting the larvae.

As the larvae grow, they will begin to resemble little

axolotls. Their exterior gills will become more prominent, and you will detect the first signs of limb development. This is an exciting period since they develop from tiny larvae to recognizable axolotls. At this point, continue to provide live food while introducing finely chopped bloodworms or earthworms. Offering a diverse range of foods ensures that they acquire all vital nutrients.

When the larvae reach about an inch in length, they will be more robust and able to handle slightly larger food pieces. You can begin to incorporate high-quality axolotl pellets into your diet. This gradual shift allows children to adjust to diverse types of food and prepares them for the diet they will follow as adults.

During this phase of rapid growth, the larvae should be regularly monitored for symptoms of stress or disease. Healthy larvae should be active and have a strong feeding response. Any indicators of lethargy, loss of appetite, or unusual swimming activity may suggest water quality issues or health problems. Regularly checking the water and making required modifications helps to avoid these concerns.

As the larvae develop, they will require more space. Overcrowding can cause competition for food and excessive waste production, stressing the larvae and impairing their growth. If you have many larvae, consider splitting them into multiple tanks to give each room to thrive. This also makes it easier to monitor water quality and feeding.

When the larvae reach around 2 inches in length, they are called juveniles and can be cared for similarly to adult axolotls. Their diet should consist of live food, chopped earthworms, and pellets. Please continue to provide regular water changes and regularly monitor their health. At this point, you can introduce them to a more regimented feeding schedule, delivering food once or twice daily.

Patience is essential for growing axolotl larvae. Each larva will grow at its own pace; thus, meeting its specific requirements is critical. Some people develop faster than others, which is quite normal. Providing constant care and a stable environment increases the likelihood that your larvae will grow into healthy adults.

Keep a close eye on their limbs and gills as they develop.

Proper limb growth is a good predictor of overall health, and colorful, fluffy gills suggest adequate oxygenation and water quality. Any anomalies, such as missing limbs or deformed gills, should be corrected immediately. These problems can sometimes be resolved with diet or water changes, but veterinarian advice is required in other situations.

Raising axolotl larvae is a gratifying activity that provides a unique look into the lives of these intriguing creatures. You may help them grow and thrive at every stage by providing a clean, stable environment and a varied, healthy diet. Seeing how they grow from tiny larvae to strong juveniles demonstrates how much care and commitment you put into their health. With careful attention and adequate care, your axolotl larvae will develop into healthy, thriving adults, ready to continue the life cycle independently.

Chapter 8

ADVANCED CARE TECHNIQUES

AQUASCAPING FOR AXOLOTLS

DESIGNING A NATURALISTIC TANK

Creating a naturalistic tank for axolotls, also known as aquascaping, blends aesthetics and utility to create a home that resembles their native environment. This procedure requires intelligent design, material selection, and a thorough understanding of the needs of these unusual amphibians. A well-designed naturalistic tank looks good and helps the axolotls' health and well-being.

Begin with the substrate, which will serve as the basis for your aquascape. Axolotls prefer fine sand. It stimulates the soft, muddy bottoms of their native environments, allowing them to burrow and forage without the risk of impaction that gravel might cause. Spread a layer of fine sand evenly around the tank bottom, ensuring it is deep enough to cover the base and allow plant roots to attach if you use live plants.

Next, evaluate the tank's layout and structure. Axolotls thrive in surroundings with plenty of hiding nooks and shaded regions. Use hardscape features such as driftwood, rocks, and caverns to do this. Driftwood provides cover and releases tannins into the water, resulting in a more natural and somewhat acidic environment. To avoid harm, choose pieces of driftwood with smooth surfaces and no sharp edges. To secure the driftwood, partially bury it in the substrate or anchor it with pebbles.

Rocks are essential in aquascaping because they enhance visual appeal and provide extra hiding spots. Choose smooth, rounded stones to protect the fragile skin of axolotls. Arrange the rocks to create stable caves and overhangs that will not collapse. Stacking stones can form multi-level structures, giving the tank a more dynamic and engaging design.

Plants are necessary in a naturalistic axolotl aquarium. They offer shelter, improve water quality by absorbing nitrates, and enhance the overall appearance. Java fern, anubias, and water wisteria are ideal alternatives since they flourish in low-light, cool-water environments that axolotls enjoy. Java fern and Anubis can be tied to driftwood or rocks with fishing lines or aquatic plant glue for a more natural appearance. Water wisteria can be planted directly in the substrate, where it will grow and spread, creating dense foliage for axolotls to hide in.

Floating plants, such as duckweed or water lettuce, can offer another level of complexity to the aquascape. These plants help disperse light, producing shady regions and relieving stress on the axolotls. They also compete with algae for nutrients, which helps keep the tank clean.

Lighting is an essential component of aquascaping. Axolotls love dim areas, so they use muted lighting that reflects their native habitat. L.E.D. lights are an excellent choice because they are energy efficient and generate little heat. Consider employing adjustable-intensity lights to fine-tune the lighting. Position the lights to create a gradient of brightness, with some parts brighter than others, which will increase the tank's apparent depth.

Filtration is another critical component. Axolotls generate a lot of waste, so an efficient filtration system is required to maintain water quality. Axolotls enjoy tranquil waters, so a canister or sponge filter with slow flow is excellent. The system should use mechanical and biological filtration to eliminate dirt and promote good bacteria growth. Position the filter input and outflow to reduce strong currents and promote uniform circulation throughout the tank.

Aquascaping also includes the careful placement of other materials, such as leaf litter and mosses. Leaf litter, such as dried Indian almond leaves, can be strewn throughout the substrate to give it a more natural appearance while releasing significant tannins into the water. Aquatic mosses, such as java moss, can be glued to rocks and driftwood to soften the hardscape while offering extra hiding locations for axolotls.

Temperature control is essential in a naturalistic axolotl aquarium. These animals thrive in cooler water, ideally 60 to 68 degrees Fahrenheit. Use an aquarium chiller to keep the temperature stable, particularly in warmer locations. Consistent temperature regulation reduces stress and encourages healthy behavior.

Regular care is required to maintain the aquascape's health. To preserve water quality:

Perform 20-30% water changes weekly.

Prune plants as needed to prevent overgrowth and ensure they do not block too much sunlight.

Check the stability of rocks and driftwood regularly to avoid shifts that could injure the axolotls.

Observe your axolotls in their new surroundings. Monitor how they engage with the various parts and make necessary changes to improve their comfort and excitement. An ideal naturalistic tank should promote natural activities, including foraging, hiding, and exploring.

Creating a naturalistic tank for axolotls requires a combination of aesthetic and functional factors. You can create a beautiful and practical environment that promotes your axolotls' health and happiness by carefully selecting and organizing substrate, hardscape pieces, plants, and other features. This careful approach to aquascaping benefits the axolotls and creates a visually gorgeous exhibit that everyone may enjoy.

ADVANCED PLANT CARE

Advanced plant care in an axolotl tank necessitates a deep awareness of the unique requirements of both the plants and the axolotls. Keeping a thriving, planted aquarium requires careful plant selection, optimal lighting, fertilizer availability, and adequate water management.

Begin by selecting plant species that thrive in the chilly, low-light environment preferred by axolotls. Java fern, anubias, and water wisteria are ideal candidates because of their hardiness and low light requirements. Java fern and Anubis can be affixed to driftwood or rocks to give the environment a natural appearance while providing cover for the axolotl. These plants grow slowly and do not require high lighting, making them suitable for an axolotl aquarium. On the other hand, maybe water wisteria is planted directly in the substrate and develops quickly, providing dense foliage for your axolotls to explore and hide under.

Lighting is essential for plant growth, but it must be balanced to benefit both plants and axolotls. Axolotls enjoy dark lighting; therefore, they use LED lights that provide a mild spectrum suitable for plant development without overwhelming the tank's residents. A timer can help regulate the light cycle, offering approximately 8-10 hours of light every day. This equilibrium ensures the plants receive enough light to photosynthesize while the axolotls remain comfortable.

Nutrient availability is another important consideration in advanced plant care. Plants in a naturalistic axolotl tank get nutrients from the water column and substrate. A nutrient-rich substrate can directly provide vital minerals and elements to plant roots. Furthermore, liquid fertilizers, especially for aquariums, can be added to the water to provide nutrients, including nitrogen, phosphorous, potassium, and trace elements. When applying fertilizer, it is critical to carefully follow the manufacturer's directions to avoid nutritional imbalances that could harm the axolotls.

Carbon dioxide (CO_2) supplementation can significantly improve plant growth, particularly in densely planted tanks. However, axolotls require stable water conditions; therefore, CO_2 levels must be carefully monitored to avoid pH swings. A CO_2 injection system with a dependable regulator and diffuser can aid in maintaining stable CO_2 levels. PH and CO_2 concentrations are regularly monitored to ensure a balanced environment for plants and axolotls.

Pruning and upkeep are vital for keeping plants healthy and preventing overgrowth. Trim dead or damaged leaves regularly to promote new growth and avoid decay, which can impact the water quality. Trimming fast-growing species such as water wisteria periodically helps preserve the correct shape and keeps the plants from overshadowing others in the tank. Furthermore, trimming promotes bushier vegetation, which provides additional hiding places for axolotls.

Managing water conditions is essential for advanced plant care. Regular water changes help to maintain adequate nutrition levels and prevent the accumulation of hazardous chemicals. Aim for 20-30% weekly water changes to keep the water clean and well-oxygenated. To maintain stable conditions, regularly test the water for ammonia, nitrites, nitrates, and pH. Ammonia and nitrite levels should be nil, with nitrates at 20 ppm. The pH should be maintained between 6.5 and 8.0, ideal for plants and axolotls.

Controlling algae is another critical component of keeping an axolotl tank planted. Algae can compete with plants for nutrients and light, resulting in poor plant health. Properly balance the light cycle and nutrition levels to manage algae growth effectively. Avoid overfeeding the axolotls since uneaten food can contribute to nutrient accumulation, encouraging algae growth. Introducing algae-eating creatures, like snails or shrimp, can also assist in managing algae, as long as they are compatible with the axolotl.

Another advanced plant care technique is root tabs, nutrient-rich tablets embedded in the substrate near the plant roots. These tabs steadily release nutrients, ensuring a consistent plant supply and minimal impact on the water column. Root tabs are handy for heavy root feeders, such as swords and crypts.

Regular observation and modification are essential for successful advanced plant care. Pay attention to the plant's growth patterns and overall health, and modify the lighting, fertilization, and water settings as needed. Each tank is unique, and achieving the ideal balance necessitates time and experimentation.

Using advanced plant care procedures, you can create a rich, thriving habitat for the plants and the axolotls. The end product is a gorgeous, lifelike aquarium that enriches the axolotls and improves the aesthetics of your home.

DIY PROJECTS FOR AXOLOTL ENTHUSIASTS

BUILDING CUSTOM HIDES AND DÉCOR

For axolotl fans, making unique hides and décor may be a satisfying hobby that adds a personal touch to your pet's surroundings. These projects not only improve the tank's aesthetics but also provide necessary hiding places and enrichment for your axolotls. With some simple resources and a little creativity, you may design and build unique features to meet the particular demands of your axolotls.

Custom hides are one of the most accessible but effective D.I.Y. crafts. Axolotls require a variety of hiding places to feel safe because they are inherently shy and seek refuge from bright lights and potential stressors. Gather aquarium-safe items like P.V.C. pipes, terracotta pots, and natural slate. These materials are strong, non-toxic, and widely available.

P.V.C. pipes are versatile and straightforward to use. They can be cut into different lengths and shapes with a hacksaw or a P.V.C. cutter. Sand down the edges to ensure they are smooth and safe for your axolotls. Coat the PVC with aquarium-safe silicone to achieve a more natural appearance, and then apply gravel, sand, or moss to the surface. This hides the pipe and creates a textured surface for axolotls to investigate.

Terracotta pots are another excellent choice for hiding. Choose pots that are large enough to fit your axolotls comfortably. You can arrange them sideways in the tank or smash them to form cave-like structures. To avoid damage, use sandpaper to smooth out any rough edges. Terracotta can be adorned with moss or aquatic plants to integrate into the tank's surroundings.

Natural slate is ideal for creating customized caves and overhangs. Slate can be piled and secured using aquarium-safe silicone to provide multi-level hiding places. Your axolotls can use these constructions for shelter and climbing. Test the stability of the slate constructions outside the tank before placing them in the water to ensure they do not collapse.

In addition to hides, you may make one-of-a-kind décor items for your axolotl tank that provide appeal and practicality. Driftwood is a popular aquarium décor item due to its natural appearance and adaptability. Large pieces of driftwood can serve as focal points in the tank, while smaller pieces can be placed to create complicated structures. Soaking driftwood before putting it in the tank keeps it from floating and releases tannins, which can result in more natural water chemistry.

Another D.I.Y. project is to create floating islands or platforms. These provide axolotls with resting areas closer to the water's surface, which they occasionally use. To make a floating island, start with a piece of Styrofoam. Cut it to the desired form and cover it with aquarium-safe silicone. To give the surface a more natural appearance, press sand, gravel, or small stones onto it. Attach aquatic plants to the island with plant weights or ties. The floating island can be secured to the tank's bottom using a fishing line and a little weight, allowing it to move slightly with the water currents.

Consider building a feeding station to provide a more participatory component. This keeps the tank clean by keeping food from dispersing and allows you to watch your axolotl's feeding habits easily. A shallow dish or container can build a rudimentary feeding station. Attach suction cups to the bottom of the dish and set it in the tank's designated feeding area. You can even create a more complicated feeding platform out of slate or acrylic, providing a stable and elevated surface for feeding.

A planted background is another D.I.Y. project that improves the tank's beauty while providing additional shade for your axolotls. Use a plastic or mesh panel as the base, then connect live plants like java fern, anubias, or moss with fishing lines or plant glue. Place the panel against the back of the tank, where the plants will eventually grow and cover the surface, creating a rich, green backdrop.

Incorporating live plants into your D.I.Y. projects not only improves the aesthetics of the tank but also benefits the general health of the ecosystem. Plants improve water quality by absorbing nitrates and supplying oxygen. Axolotls need chilly water temperatures, so choose robust, low-light plants.

When constructing unique hides and décor, always prioritize your axolotls' safety and well-being. Avoid utilizing objects that may leach hazardous substances into the water or have sharp edges that could cause injury. Before placing any D.I.Y. objects in the tank, ensure they are stable and durable. After adding new objects, closely monitor your axolotls to verify they are adjusting well and that the new additions do not cause stress or harm.

Creating bespoke hides and décor for your axolotl tank is a

rewarding method to improve the habitat while displaying creativity. Use safe materials and thoughtful design to create a stimulating and secure environment that enhances your axolotls' health and pleasure. Whether it's a simple P.V.C. hide, a genuine slate cave, or a floating island, these D.I.Y. projects add unique elements that enrich your axolotl's existence while making your aquarium aesthetically pleasing and fascinating.

HOW TO CONSTRUCT CUSTOM HIDES

Creating custom hides for axolotls is a rewarding project that enhances their tank environment and ensures they feel secure. Here's a detailed guide on the materials you'll need and step-by-step instructions on how to construct different types of custom hides.

Materials Needed:

1. PVC Pipes:

   - PVC pipes (various diameters)

   - Hacksaw or PVC cutter

- Sandpaper

- Aquarium-safe silicone

- Gravel, sand, or moss (optional for decoration)

2. Terracotta Pots:

   - Terracotta pots (various sizes)

   - Hammer or pliers (for breaking pots)

   - Sandpaper

   - Moss or aquatic plants (optional for decoration)

3. Natural Slate:

   - Slate pieces

   - Aquarium-safe silicone

   - Sandpaper

4. Driftwood:

   - Driftwood pieces

- Bucket for soaking

- Sandpaper (if needed)

5. Styrofoam (for floating hides):

   - Styrofoam sheets

   - Aquarium-safe silicone

   - Sand, gravel, or small stones (for decoration)

   - Fishing line

   - Small weights

Step-by-step instructions:

1. P.V.C. Pipe Hides:

Cut the Pipe: A hacksaw or P.V.C. cutter can measure and cut the P.V.C. pipe to the desired length. Standard lengths range from 6 to 12 inches.

Smoothing Edges: After cutting, use sandpaper to smooth down any rough edges to avoid injuring your axolotl.

Decorating is optional. Apply a thin layer of aquarium silicone to the pipe's outer surface. To give the silicone a more natural appearance, press gravel, sand, or moss onto it. Allow it to dry thoroughly before putting it into the tank.

2. Terracotta Pot Hides:

Prepare the pot by placing it sideways in the tank. Carefully crack the pot with a hammer or pliers to create a more cave-like structure.

Smoothing Edges: Sand down any rough edges to avoid harm.

Positioning and decoration: Place the pot or pot pieces within the tank. Adorn them with moss or attach aquatic plants with fishing lines or glue to give them a more natural appearance.

3. Natural Slate Caverns:

Selecting Pieces: Choose a flat slate that is easy to stack.

Constructing the Cave: Arrange the slate pieces to create a cave or overhang. Use aquarium-safe silicone to hold the pieces together, ensuring the structure remains solid.

Smoothing Edges: Sand any sharp edges to avoid injury.

Stability Check: Test the structure's stability outside the tank. Adjust as needed before putting it in the aquarium.

4. Driftwood Hides:

Preparation: Soak driftwood in a bucket of water for days to prevent floating and release tannins.

Placement: Arrange the driftwood in the tank to make hiding places and overhangs. Large pieces can be used as focal points, while smaller pieces can be piled or arranged strategically.

Smoothing Edges: Sand down any rough or sharp edges to guarantee safety.

5. Floating Hides:

Cut Styrofoam into suitable shapes and sizes for the floating island or platform.

Applying Silicone: Spread a thin layer of aquarium-safe silicone over the Styrofoam surface.

Decorating: To achieve a natural look, press sand, gravel, or small stones onto the silicone. Let it dry completely.

Anchoring: Thread the fishing line through the corners of the Styrofoam and attach small weights to the ends. This will keep the hide in place while allowing for mobility with the water currents.

Additional Tips:

Safety First: Always ensure that all materials are non-toxic and suitable for aquarium use. Avoid using paints or adhesives that were not intended for aquatic conditions.

Testing: Before inserting any D.I.Y. hide in the tank, properly rinse it with dechlorinated water and test it outside to ensure stability.

Observation: After introducing new hides, ensure that your axolotl is comfortable and that the new additions do not cause stress or injury.

By following these procedures, you may make a variety of bespoke hides that will improve your axolotl's tank habitat, giving them areas to explore, rest, and feel safe.

HOW TO BUILD FLOATING ISLANDS OR PLATFORMS

Creating floating islands or platforms in an axolotl tank can add a unique aesthetic and provide additional resting and exploration areas for your pets. Here's a detailed guide on the materials needed and step-by-step instructions on how to construct these floating features.

Materials Needed:

1. Styrofoam Sheets:

   - Styrofoam sheets (1/2 inch to 1 inch thick)

   - Utility knife or hot wire cutter

2. Aquarium-Safe Silicone:

   - Aquarium-safe silicone sealant

   - Caulking gun (if needed)

3. Decorative Elements:

   - Fine sand, gravel, or small stones

   - Aquarium-safe moss or aquatic plants

   - Plant weights or anchors

4. Fishing Line and Weights:

   - Fishing line

- Small, non-toxic weights (e.g., stainless steel washers or aquarium weights)

5. Optional Tools:

   - Sandpaper (for smoothing edges)

   - Ruler or measuring tape

   - Permanent marker (for marking cuts)

Step-by-Step Instructions:

1. Cut the Styrofoam.

Measure and mark. Use a ruler or measuring tape to figure out the size and shape of your floating island or platform. Rectangles, squares, and irregular natural shapes are all standard shapes. Mark the outline on the Styrofoam sheet with a permanent marker.

Cut the Styrofoam: Gently cut following the lines using a utility knife or hot wire cutter. Ensure the edges are as smooth as possible to avoid jagged parts that could hurt your axolotl.

2. Smoothing the edges:

Sand the edges: If necessary, use sandpaper to smooth the edges of the Styrofoam to ensure no sharp or rough spots could damage your axolotl.

3. Use Aquarium-Safe Silicone:

Prepare the surface. To remove any dust or particles from the Styrofoam surface, wipe it down with a moist cloth.

Apply Silicone: Using aquarium-safe silicone and a caulking gun (if necessary), spread a thin, and even silicone coating across the surface of the Styrofoam.

Decorate the surface. To give the silicone-covered surface a more natural appearance, press fine sand, gravel, or small stones onto it. Ensure that the decorative components are securely embedded in the silicone. Allow the silicone to dry completely, following the manufacturer's instructions.

4. Optional: Add Plants and Moss.

Adherence Plants: If you're using live or artificial plants, use aquarium-safe silicone to adhere them to the Styrofoam. To keep live plants in place, try using plant weights or anchors.

Add Moss: Apply aquarium-safe moss to the surface with

small amounts of silicone. This can give your axolotl a more natural appearance while providing more hiding locations.

5. Anchor the Floating Island:

Prepare the Fishing Line: Cut the fishing line into lengths sufficient to anchor the island at the chosen depth in the Tank.

Attach the fishing line. Add little silicone drops to the Styrofoam's edges or corners to connect the fishing line. Ensure that the connection points are secure.

Attach Weights: Secure the other ends of the fishing line with small, non-toxic weights. These weights will help anchor the island and keep it from floating about the Tank.

6. Place the Floating Island in the Tank:

Position the island: Carefully position the floating island in the Tank. Adjust the fishing line and weights to get the island at the right height and location.

Secure the Weights: Place the weights tightly in the bottom of the Tank to keep the island steady.

Additional Tips:

Waterproofing: Use silicone to fasten any decorations or

plants stuck to the Styrofoam.

Observation: Monitor your axolotl's interactions with the new floating island to ensure they do not cause stress or harm.

Regularly inspect the floating island for signs of wear and damage. To ensure your axolotl's safety, replace or repair items as needed.

Following these procedures, you can make a functional and visually appealing floating island or platform for your axolotl. These features give additional resting and exploration places, improving the overall habitat and promoting the health of your axolotls.

DIY FILTER SYSTEMS

Creating your own D.I.Y. filter system for an axolotl tank can be rewarding, especially if the filtration system suits the specific needs of your aquatic creatures. Axolotls generate a lot of waste; therefore, having clean water is critical for their health. A well-designed D.I.Y. filter can provide mechanical, biological, and chemical filtering, ensuring the tank's cleanliness and safety.

The first step in building a D.I.Y. filtration system is understanding the fundamental components and their functions. A primary aquarium filter utilizes three types of filtration: mechanical, biological, and chemical. Mechanical filtration collects particles and waste, biological filtration uses beneficial bacteria to break down toxic ammonia and nitrites, and chemical filtration uses activated carbon or another medium to remove dissolved compounds such as smells and discoloration.

To begin, collect the necessary items. You'll need a plastic container that fits within or above your tank, like a food-safe plastic storage box or a small bucket. You'll also need a submersible water pump, filter media (sponges, ceramic rings, bio-balls, and activated carbon), tubing, and a drill or utility knife to make holes.

Start by prepping the plastic container that will hold your filter media. Drill or cut several small holes in the sides and bottom of the container to allow water to pass through. These pores should be tiny enough to keep the filter media from dropping out but large enough to allow appropriate water passage. Smooth any rough edges to avoid damage to the tubing or the tank.

Next, place the filter media in the container. Begin with a layer of coarse sponge or filter pad at the bottom for mechanical filtering. This layer will capture more considerable dirt and particles. On top of this, place a layer of biological media, such as ceramic rings or bio-balls, to allow good bacteria to colonize. Finally, add a layer of activated charcoal or another chemical filtration media to remove any dissolved contaminants from the water.

Now, it's time to hook up the water pump. Select a submersible pump appropriate for your tank's size and the flow rate needed. Connect a piece of tubing to the pump's outlet and position the pump at the bottom of the tank. Run the tube to the filter container and ensure its properly secured. The pump will extract water from the tank and pass it through the filter media, which will be cleaned before returning it to the tank.

Place the filter container above the tank or within a compartment if your tank configuration allows. Make sure the container is stable and won't tip over. If you put the container above the tank, you may need to build an essential support or shelf to keep it in place. The water from the pump should run through the filter medium and back into the tank via an overflow system or by creating a minor waterfall effect, which also aids in aeration.

To test the filter system, turn on the pump and observe the water flow. Ensure the water flows smoothly through the filter medium and back into the tank with no leaks or blockages — adjust the tubing position and the pump's flow rate to obtain the best performance.

Regular maintenance of the D.I.Y. filtration system is required to guarantee it performs correctly. To eliminate trapped dirt, regularly clean mechanical filter media (such as sponges or filter pads). Rinse the biological media with tank water during water changes to keep the beneficial bacteria. Replace the activated carbon or other chemical filtration media every few weeks since it depletes with time.

Additionally, examine the water parameters in your tank to ensure that the filter provides a healthy environment for your axolotls. Regularly test the water for ammonia, nitrites, nitrates, and pH. Make partial water changes as needed to keep these parameters within acceptable limits. The D.I.Y. filtration system should reduce the required water changes, but constant monitoring is still necessary.

You can keep your axolotl tank clean and healthy by building a D.I.Y. filter system adapted to its specific demands. This project provides a cost-effective solution and gives you the joy of creating a personalized arrangement that correctly fits your tank's needs. With appropriate maintenance and regular monitoring, your D.I.Y. filter will keep the water clean and healthy for your axolotls for many years.

Chapter 9

AXOLOTLS IN THE COMMUNITY

JOINING THE AXOLOTL COMMUNITY

ONLINE FORUMS AND GROUPS

Joining the Axolotl community may be a rewarding experience, providing a wealth of knowledge, support, and friendship with other fans. One of the most effective methods to connect with this community is through internet forums and organizations dedicated to axolotl care and admiration. These platforms provide excellent resources, from expert advice to personal experiences, making them essential for novice and experienced axolotl owners.

Online forums are an excellent starting point. These sites

allow people to ask questions, share information, and debate various aspects of axolotl care. Axolotl-specific areas on websites such as Caudata.org enable members to discuss everything from tank setup and maintenance to health issues and breeding. The forum layout keeps information organized and easily accessible, making it a valuable resource for any specific queries or issues.

One primary advantage of joining online forums is the wealth of information available. Many members are experienced axolotl keepers and breeders willing to share their knowledge. They can bring insights not commonly seen in regular pet care books or websites. For example, if your axolotl is exhibiting strange behavior or symptoms, a post on a forum can elicit answers from numerous people who have faced and handled similar problems. This common wisdom can be extremely reassuring and helpful.

Social media groups play an essential part in the axolotl community. Numerous communities for axolotl aficionados exist on platforms such as Facebook and Reddit, sharing images, videos, and advice. These groups are more casual and immediate than forums, allowing for faster interactions and updates. For example, Facebook groups like "Axolotl Owners" and Reddit's r/axolotls subreddit are active, providing a forum for rapid questions and in-depth conversations. Members of these organizations frequently share personal experiences and celebrate milestones such as their axolotl's development or successful breeding, which can foster a strong feeling of community.

Joining these groups is simple. On Facebook, you can look for axolotl-related groups and ask to join. Most groups have rules to encourage respectful and constructive interactions; new members must usually agree with them before joining. Once you've become a member, you may begin engaging in discussions, asking questions, and sharing your experiences. On Reddit, joining is as simple as subscribing to the appropriate subreddit. Participating is simple; you may ask questions, leave comments on other postings, and even upload photographs and videos of your axolotl.

Discord is another place where axolotl fans can connect. Discord servers for axolotls offer a more participatory experience, including real-time conversation and the opportunity to create various channels for different topics.

This style is ideal for developing a more personal connection with other members and providing immediate input on queries or concerns. Some servers even organize live events, such as expert-led Q&A sessions or virtual tours of member tanks.

In addition to these venues, many axolotl lovers have blogs and YouTube channels. These materials can provide detailed instructions and visual guides for issues such as tank setup, feeding, and health care. Watching videos of other axolotl owners might give you valuable tips and ideas for your tank setup. Engaging with content creators through comments and social media makes you feel more connected to the community.

The advantages of joining the axolotl community go beyond simply learning information. These groups provide emotional support and a sense of belonging. Sharing the joys and trials of axolotl care with others who understand may be extremely rewarding. Whether going through a health crisis or celebrating your axolotl's first birthday, having a community to lean on may make the experience much more enjoyable.

Furthermore, joining these groups can encourage you to share your knowledge and experiences. As you gain experience, you can guide new axolotl owners by offering ideas and information you've acquired. This cycle of learning and teaching benefits the community by ensuring that knowledge is handed along and developed.

Joining the axolotl community through online forums and organizations provides valuable resources and assistance. Whether you're looking for expert guidance, sharing your experiences, or simply enjoying the company of like-minded enthusiasts, these platforms offer a tremendous opportunity for interaction and learning. Participating in these groups improves your axolotl-keeping expertise and adds to axolotls' collective knowledge and well-being worldwide.

List of Forums And Group

Certainly! Here is a list of both physical and online forums where axolotl enthusiasts can connect, share information, and seek advice. All the listed forums are well-known and credible within the axolotl-keeping community.

Online Forums:

1. Caudata.org

Website: [Caudata.org Axolotl Forums] (https://www.caudata.org/forums/forums/axolotls.2/)

Description: A comprehensive forum dedicated to newts and salamanders, with a specific section for axolotls. It offers in-depth discussions on care, breeding, health issues, and tank setups.

2. Reddit - r/axolotls

Website: [Reddit r/axolotls] (https://www.reddit.com/r/axolotls/)

Description: A popular subreddit where axolotl owners share photos, ask questions, and discuss all aspects of axolotl care. It's very active and welcoming to new members.

3. Facebook Groups:

Axolotl Owners

Website: [Axolotl Owners Facebook Group] (https://www.facebook.com/groups/axolotlowners/)

Description: A large and active group where members share advice, photos, and experiences related to axolotl care.

Axolotl Enthusiasts

Website**: [Axolotl Enthusiasts Facebook Group] (https://www.facebook.com/groups/axolotlenthusiasts/)

Description: Another popular group with a strong community focus, offering support and knowledge-sharing for axolotl keepers.

4. Axolotl Forum

Website: [Axolotl Forum] (https://www.axolotl.org/forum/)

Description: A dedicated forum for axolotl enthusiasts, covering a wide range of topics from beginner tips to advanced breeding techniques.

5. Discord Servers:

Axolotl Chat

Website: Invite link usually shared within community platforms like Reddit or Facebook.

Description: Real-time chat and discussion server for axolotl keepers, offering instant advice and a vibrant community atmosphere.

Physical Forums (Clubs and Societies):

1. Local Aquarium Societies:

Many regions have local aquarium or herpetology societies that meet regularly and often have members who keep axolotls.

How to Find: Search for "aquarium society" or "herpetology club" along with your city or region. Examples include the "Greater Seattle Aquarium Society" or the "New York Turtle and Tortoise Society".

2. Reptile and Amphibian Expos:

Events where reptile and amphibian enthusiasts gather. These expos often feature breeders and vendors specializing in axolotls.

How to Find: Look up reptile expos or fairs in your area, such as the "Repticon" series in the United States.

3. Zoos and Aquariums with Amphibian Programs:

Some larger zoos and aquariums have specific programs or days dedicated to amphibians, where you can meet experts and other enthusiasts.

Examples: The "San Diego Zoo", "Smithsonian National Zoological Park", and the "Aquarium of the Pacific" often have educational events and exhibits on amphibians.

4. Herpetological Societies:

Example: The "British Herpetological Society" (BHS) holds meetings and events for reptile and amphibian enthusiasts, which can be an excellent opportunity to connect with others interested in axolotls.

Website: [British Herpetological Society] (https://www.thebhs.org/)

5. Pet Stores with Aquatic Departments:

Some local pet stores host workshops or meetups for fish and amphibian keepers.

How to Find: Check with larger chain stores like "Petco" or "Petsmart", or local specialty stores for event schedules.

By joining these online and physical forums, you can access a wealth of information, connect with other axolotl enthusiasts, and become a part of the broader community dedicated to the care and appreciation of these fascinating creatures.

ATTENDING AXOLOTL EXPOS AND EVENTS

Attending axolotl expos and events provides an engaging experience for any enthusiast, allowing them to learn, network, and immerse themselves in the world of these fascinating creatures. These events are more than just displays; they are centers of knowledge and community, bringing together breeders, professionals, amateurs, and interested newcomers.

Consider strolling into an expo hall filled with excitement. The air is filled with discussions about tank configurations, feeding procedures, and the latest axolotl care technology. These events offer a variety of knowledge and products all in one location. From unusual morphs to high-quality food sources, the range is remarkable. Vendors display a variety of axolotls, with each tank meticulously set up to highlight the healthy, vivid animals. It's a visual feast and a treasure trove for anyone trying to improve their axolotl habitat.

One of the most significant advantages of attending these events is meeting and interacting with breeders and specialists. These folks often have years, if not decades, of expertise caring for and breeding axolotls. They are frequently eager to share their knowledge, answer questions, and offer insights you won't find online or in books. Conversations with these experts will help you learn about recent research, breeding techniques, and care improvements that can benefit your practices.

Workshops and seminars are another hallmark of Axolotl Expos. These seminars cover various topics, from fundamental care and tank setup to advanced breeding and disease control. Participating in these classes can help you better understand axolotl biology and care. For example, you may attend a session on the genetics of axolotl color morphs to discover how different features are passed down and how to care for unique types. Hands-on courses may teach you how to make unique tank decorations or set up a bioactive terrarium, skills you can immediately apply to improve your axolotl's environment.

Attending these events also provides essential networking opportunities. Meeting other enthusiasts face to person allows you to share experiences, swap stories, and form relationships within the community. These partnerships can be quite beneficial, offering continuing support and advice. Whether solving a specific issue or looking for moral support, having a network of other axolotl enthusiasts is essential.

Many expos include competitions and showcases. These can include best-in-show competitions for the most beautiful or distinctive axolotls and aquascaping events that showcase innovative and effective tank layouts. Entering these contests can be a fun way to showcase your axolotl and tank design abilities, and winning can provide a sense of accomplishment and acknowledgment within the community.

Shopping is a big attraction at these gatherings. Expos frequently feature vendors offering specialized equipment, food, and decorations unavailable in traditional pet stores. You may stock up on high-quality materials, discover new products, and even get uncommon axolotl morphs for your collection. Because you can see and handle things in person, you can make better decisions than when shopping online.

Furthermore, attending these gatherings might provide insight into the broader trends and challenges affecting the axolotl community. There are frequent discussions regarding conservation, breeding ethics, and the most recent scientific breakthroughs. Participating in these discussions allows you to stay informed and engaged on the most important subjects for the community.

These events are precious to people who are new to axolotl

keeping. They provide a complete introduction to the activity, serving as a one-stop shop for learning, shopping, and socializing. Seeing a range of axolotls and setups in person may be exceptionally motivating, allowing you to imagine your axolotl care possibilities.

Planning a trip to an expo or event requires some preparation. Check the event schedule beforehand to determine which workshops, seminars, and vendors you wish to attend. Arriving early allows you to avoid the crowds and make the most of your trip. Bring a notebook to jot down ideas and advice, and remember to take photos of the various setups and axolotls on the show.

Attending axolotl expos and events is more than a day out; it's an opportunity to expand your knowledge, improve your methods, and integrate more fully into the axolotl community. Whether you're an experienced keeper or a curious novice, these events have something for everyone. By engaging, you enrich your individual experience while contributing to the collective knowledge and passion propelling the community forward.

CONSERVATION EFFORTS

SUPPORTING WILD AXOLOTL POPULATIONS

Conservation initiatives to protect wild axolotl populations are critical to averting the extinction of these unique creatures. Wild axolotls, native to Mexico's ancient lake complex of Xochimilco, are threatened by various factors, including habitat loss, pollution, and invasive species. Participating in and supporting conservation efforts can significantly impact maintaining this unique species for future generations.

The fate of the wild axolotl has become a metaphor for the more significant environmental concerns confronting the Xochimilco region. Xochimilco was formerly an extensive network of beautiful shallow lakes and canals, but development has severely transformed the area. Mexico City's growth and continued expansion have resulted in severe habitat degradation. Wetlands have been drained, water quality has worsened owing to pollution, and the introduction of non-native species like tilapia and carp has exacerbated the situation. These invasive fish compete with axolotls for food and feed on their eggs and larvae, putting the wild axolotl population at risk.

Habitat restoration is frequently the first step in supporting wild axolotl populations. Restoring the natural ecosystem of Xochimilco entails reestablishing wetlands, restoring water quality, and eradicating exotic species. These efforts are collaborative efforts between local and international conservation organizations. They create artificial refuges within the canals to provide safe breeding grounds and shelter for axolotls. These refuges are meticulously engineered to simulate natural circumstances, protecting the axolotls from predators and pollution.

Public awareness and education are critical components of conservation initiatives. Many people need to be aware of the axolotl's threatened condition or the ecological significance of Xochimilco. Educational programs aim to educate the local community and the general public about the issues that axolotls face and how they might be helped. Schools and universities frequently collaborate with conservation organizations to provide hands-on learning opportunities for students to participate in habitat restoration initiatives and scientific research. These activities build a sense of responsibility and inspire future generations to continue conservation efforts.

Research is vital to conservation. To establish successful conservation strategies, scientists investigate axolotls' biology, behavior, and ecology. Captive breeding programs are critical for preserving genetic variation and developing a stable population that can be returned to nature. These initiatives are painstakingly controlled to guarantee that captive-bred axolotls retain the genetic traits required for survival in their natural environment. Researchers also monitor wild populations to evaluate the efficacy of conservation efforts and make necessary changes.

Financial support for these projects is another way to help. Donations to conservation groups support essential programs ranging from habitat restoration to public education campaigns. Many groups, such as the Mexican Axolotl Conservation Project and international agencies like the World Wildlife Fund, rely on public funding to continue their work. Even small donations can significantly impact purchasing equipment, funding research, and supporting local conservationists.

Individuals can take direct action to help wild axolotls. Volunteer programs, whether local or as part of overseas conservation tours, allow people to provide hands-on assistance. Volunteers may assist with creating artificial refuges, eliminating invasive species, and performing population surveys. These experiences help us understand the obstacles that wild axolotls confront and the actions required to safeguard them.

Promoting sustainable habits in daily life also helps with conservation. Reducing the amount of plastic and chemicals that wind up in waterways improves water quality in natural environments. Supporting policies and activities to maintain wetlands and other vital ecosystems can bring broader environmental benefits. Please encourage others to be sensitive to their ecological impact, which contributes to developing a conservation culture beyond the axolotl.

Another part of axolotl conservation is campaigning for more environmental legislation and enforcement. Long-term conservation success depends on upholding regulations safeguarding natural areas and endangered species. Advocacy can include:

Writing to government officials.

Joining local conservation groups.

Raising public understanding about the value of environmental protection.

Conservation initiatives to protect wild axolotl populations include various activities, including habitat restoration, public education, scientific study, and advocacy. Individuals and organizations participating in these efforts can help assure the survival of this extraordinary species. Every action, whether through financial support, volunteer effort, or sustainable living practices, helps to achieve the larger objective of protecting the wild axolotl and the ecosystem in which it lives.

HOW TO GET INVOLVED

Participating in axolotl conservation and the more extensive network of enthusiasts may be both enjoyable and beneficial. There are countless ways to help these exciting creatures, whether you are a beginner or a seasoned expert. Your participation can significantly impact local projects and worldwide conservation efforts.

One of the most accessible ways to participate is through education and awareness. Begin by learning everything you can about axolotls, their natural environment, and the obstacles they confront. This knowledge broadens your comprehension and enables you to educate others. Sharing information on social media, creating blog entries, and speaking at local schools and community centers are all great ways to promote awareness. Participating in axolotl-specific online communities, such as forums and social media groups, can also help distribute helpful information and establish a supporting network.

Volunteering with local organizations dedicated to amphibian conservation is another excellent way to contribute. Many cities feature wildlife rehabilitation centers, zoos, and aquariums where volunteers can assist with educational programs, habitat restoration projects, and animal care. These organizations frequently require extra help cleaning tanks, making food, and assisting with public outreach events. Volunteering provides practical assistance, an inside view into conservation efforts, and the opportunity to network with like-minded people.

Participating in citizen science programs is an effective method to promote axolotl research and conservation. These initiatives frequently include monitoring local water bodies for amphibian populations, gathering data on water quality, and reporting sightings of wild axolotls or other native species. Websites and apps like iNaturalist allow users to record and share findings, contributing to more extensive scientific investigations. You contribute to a more thorough understanding of axolotl populations and habitats, which is critical for effective conservation management.

Financial support is also essential for many conservation efforts. Donations to organizations such as the Mexican Axolotl Conservation Project, World Wildlife Fund, and others support imperative research, habitat restoration, and public education initiatives. Many of these organizations provide membership plans that include regular updates on their activities, exclusive content, and ways to get more involved with their initiatives. Even modest gifts may make a big difference as they build up to provide critical resources for continuing conservation initiatives.

Another method to get engaged is to participate in or organize local clean-ups. Pollution poses a significant threat to axolotl ecosystems, particularly in urban places such as Mexico City, where their natural environment is constantly under strain. Joining or forming a community clean-up group can help reduce pollution in local waterways and improve ecosystem health. These events can also educate the public on keeping ecosystems clean and healthy for all wildlife.

Advocacy and activism are effective techniques for enacting more significant changes that benefit axolotls and other endangered species. Policies that conserve wetlands, limit pollution, and promote biodiversity are critical to long-term conservation success. Contacting local and national politicians, attending public hearings, and joining conservation advocacy groups can help you get your voice heard and influence decision-making. Social media campaigns and petitions are equally excellent strategies to increase public awareness and support for essential conservation actions.

Another way to become involved is by participating in captive breeding initiatives. Many zoos and aquariums run breeding programs to preserve endangered animals' genetic diversity, such as axolotls. These programs frequently recruit volunteers to assist with animal care and facility maintenance. Furthermore, some private breeders are committed to maintaining healthy, genetically diverse axolotl populations and may be open to aid or partnership. Supporting ethical breeders by purchasing from them rather than less reputable sources contributes to the health and sustainability of captive populations.

Finally, becoming an advocate for axolotls in your daily life can have a considerable impact. This might include basic steps such as using environmentally friendly items to lower your carbon footprint, advocating sustainable behaviors among friends and family, and choosing to support businesses that value environmental responsibility. Leading by example encourages others to think about their actions' environmental impact and helps foster a conservation culture.

Participating in axolotl conservation and the larger community provides several opportunities to make a difference. Education, volunteering, citizen science, financial assistance, advocacy, captive breeding initiatives, and everyday acts can all help secure these unique creatures' survival. By supporting these programs, you join a global effort to safeguard and preserve axolotls for future generations.

CONCLUSION

FINAL TIPS FOR SUCCESSFUL AXOLOTL KEEPING

RECAP OF KEY POINTS

Keeping axolotls successfully demands attention to detail, consistency, and a thorough grasp of their requirements. Whether you're a new owner or have been caring for these exciting creatures for a while, here are some last pointers to help your axolotls thrive in their watery environment.

The first and most important consideration is water quality. Axolotls are sensitive to water conditions, so maintaining a high water quality is critical. Regularly test your water for ammonia, nitrites, nitrates, and pH. Aim for zero ammonia and nitrites, nitrates below 20 ppm, and a pH of 6.5 to 8.0.

Weekly 20-30% water changes are required to keep the water clean and free of contaminants. A high-quality water conditioner can also help neutralize the chlorine and chloramines in tap water.

Temperature control is another essential part of axolotl care. These amphibians prefer cooler water, between 60 and 68 degrees Fahrenheit. Higher temperatures can increase stress and susceptibility to disease. To maintain the proper temperature, use an aquarium chiller or fans, particularly during the warmer months. Keep the tank in a cool, dark part of your home to help control the temperature naturally.

Providing a large, well-structured tank is critical for your axolotl's physical and mental health. Each axolotl requires at least a 20-gallon tank, with an additional 10 gallons for each additional axolotl. To avoid escapes, ensure that the tank's lid is secure. Create an exciting environment with plenty of hiding places using caverns, P.V.C. pipes, and aquatic plants. These factors resemble their natural environment, lowering stress and encouraging natural behaviors such as exploring and hiding.

A balanced diet is vital for your axolotl's health. Provide a range of feeds, including high-quality pellets, earthworms, bloodworms, and occasional treats such as brine shrimp. Axolotls are carnivores and require a protein-rich diet. Feed them in suitable portions two to three times weekly, and remove any uneaten food to avoid water quality issues. Monitoring their feeding habits allows them to change their diet to maintain a healthy weight and avoid overfeeding.

Axolotls prefer dim areas. Therefore, lighting should be kept to a minimum. Excessive lighting can be stressful and unhealthy. Use low-intensity L.E.D. lighting and create shaded regions throughout the tank. Floating plants and properly placed ornaments can help create a balanced light environment appropriate for your axolotl's preferences.

Regular health exams are necessary to detect any potential problems early. Check your axolotls for sickness symptoms, such as changes in appetite, strange swimming behavior, skin lesions, or gill abnormalities. If you detect any of these symptoms, contact a veterinarian with amphibian's experience. Keeping a log of their health, behavior, and feeding schedule can help diagnose and manage any issues.

Tankmates should be picked with discretion. While some axolotls can coexist happily with others, they must be regularly monitored for aggressive behavior. Axolotls' skin and gills are sensitive and readily damaged during battles. If you keep numerous axolotls together, ensure the tank is large enough to accommodate each individual and has enough hiding locations to reduce territorial behavior.

Maintaining a consistent and peaceful environment is critical. Axolotls can be stressed by unexpected changes in their environment, such as water temperature or quality changes, loud noises, or excessive handling. Establish a feeding, cleaning, and tank maintenance schedule to ensure a steady and predictable habitat. To alleviate stress, handle your axolotl cautiously and at a minimum frequency.

Research and continual learning are essential in axolotl care. Reading books, joining online forums, and engaging in community groups will update you on the most recent information and improvements in axolotl husbandry. Sharing your experiences and learning from other axolotl lovers will help you get fresh insights and improve your care techniques.

Enjoy the experience of caring for your axolotls. These strange species provide an intriguing peek into the world of amphibians. Observing their acts and conversations and assuring their well-being may be highly fulfilling. By following these guidelines and paying attention to their needs, you can foster a thriving environment where your axolotls can live long, healthy, and happy lives.

Axolotl care requires rigorous attention to water quality, temperature control, tank layout, nutrition, and health monitoring. By creating a stable and enriching environment for your axolotls, you ensure they thrive and continue bringing delight and intrigue into your life.

- Encouragement for both new and experienced owners

Owning an axolotl, whether new to the hobby or a seasoned enthusiast, can be both difficult and rewarding. These unique amphibians demand special care, but learning and adjusting to their needs is a wonderfully gratifying experience. Let's look at some encouragement for novice and seasoned axolotl owners, celebrating the successes and addressing the issues.

For new proprietors, the initial phase might be

overwhelming. There is much to learn about tank layout, water quality, feeding, and overall maintenance. It's crucial to remember that every expert started as a beginner. Mistakes are part of the learning process; each difficulty you overcome allows you to become a better caregiver. Begin with the basics:

1. Make sure your tank is cycled adequately before introducing your axolotl.

2. Keep the water clean.

3. Feed them a varied food.

4. Feel free to seek help from more experienced keepers, internet forums, or trusted sources. The axolotl community is open and supportive, always willing to exchange expertise and tips.

It's pretty satisfying to watch your axolotl thrive. Observing their unusual actions, from delicate swimming to inquisitive investigation of their surroundings, provides a sense of accomplishment. Every day presents a new opportunity to bond with your pet, learn about their preferences, and improve your care techniques. As you get to know your axolotl, you'll discover tiny signals of their well-being, allowing you to respond proactively to their requirements.

For seasoned entrepreneurs, the path continues with continual learning and discovery. Even after years of experience, there is always something new to discover. Advances in axolotl care, breeding techniques, and tank developments can help you improve your skills and keep your interest alive. Sharing your expertise with newcomers benefits them and reinforces your understanding and devotion. Teaching others and seeing them achieve provides a tremendous sense of accomplishment while strengthening the community.

Providing a healthy and enriching habitat for your axolotl is an ongoing endeavor. Regularly upgrading and improving your tank configuration ensures that it remains dynamic and exciting for you and your pet. Experimenting with various plants, hides, and decorations can help to create a stimulating environment that promotes natural behaviors. Each alteration and enhancement you make improves your axolotl's overall health and happiness.

Dealing with health difficulties or unanticipated hurdles is part of every owner's experience. It's easy to become discouraged when your axolotl has health issues, or things don't go as planned. Remember, perseverance and dedication are essential. Seek advice from vets who have worked with amphibians, and don't be reluctant to ask for support from others. Each struggle you face improves your bond with your axolotl and increases your experience and resilience as a parent.

The relationship between an owner and their axolotl is special. With their wide-eyed looks and excellent regenerating capacities, these organisms arouse astonishment and intrigue. Your patience and care are rewarded by the satisfaction of watching them thrive in a well-kept home. Every minute, from feeding time to witnessing their nocturnal activity, strengthens your bond and appreciation for these remarkable creatures.

Continuing education is essential for new and experienced business owners alike. Continue to be curious and explore new knowledge and techniques. Attend axolotl expos, workshops, and local/online groups. Engaging with other enthusiasts broadens your knowledge and gives you a sense of belonging and shared enthusiasm. The community's collective wisdom and support are vital tools to help you succeed as an axolotl keeper.

Celebrate the tiny achievements and daily joys. Whether it's successfully putting up a new tank, nursing a sick axolotl back to health, or simply watching your axolotl interact with its surroundings, these moments capture the essence of the experience. Each step you take, each improvement you make, and each observation you document contribute to a happy and enjoyable trip.

Owning an axolotl is an exciting and gratifying experience for novice and seasoned keepers. Accept the learning curve, cherish the successes, and confront the problems with tenacity. Your dedication and care build a thriving environment for your axolotl, resulting in a unique and satisfying friendship that strengthens with time. Whether you're just starting or have years of experience, the path with your axolotl is constant discovery, growth, and shared wonder.

APPENDICES

GLOSSARY OF TERMS

Creating a comprehensive glossary of words for axolotl care can be quite beneficial to both new and experienced hobbyists. This resource will clarify technical terms and concepts, allowing you to better comprehend axolotl husbandry's complexities. Here is a comprehensive vocabulary of words relating to axolotls and their care.

Ammonia (NH_3/NH_4^+)

Is a poisonous chemical created by decomposing organic materials, including uneaten food and garbage. Ammonia is toxic to aquarium axolotls and should be kept at zero levels through frequent water changes and effective biological filtration.

Aquarium chiller

Is a device that lowers the temperature of aquarium water.

Because axolotls prefer cooler water, an aquarium chiller can help maintain the proper temperature range, particularly in warmer locations.

Biological Filtration

A process in which beneficial bacteria convert poisonous ammonia and nitrites into less damaging nitrates. This form of filtration is critical for ensuring that axolotls thrive.

Bio-Balls

Are plastic spheres used for biological filtration systems? They give a vast surface area for good bacteria to colonize, improving the aquarium's breakdown of ammonia and nitrates.

Brine Shrimp

Axolotls frequently eat small aquatic crustaceans, especially when they are larvae. Brine shrimp are high in protein and easy to grow at home.

Canister Filter

An external filter that performs mechanical, chemical, and biological filtration. Canister filters are highly efficient and can manage more significant amounts of water, making them excellent for axolotl aquariums.

Cycling

Is the process of growing beneficial bacteria in aquariums to convert ammonia and nitrites into nitrates. Cycling is required before introducing axolotls to ensure water safety and stability.

Dechlorinator

A chemical additive that neutralizes chlorine and chloramines in tap water. Chlorine is toxic to axolotls, so use a de-chlorinator when changing the water.

Erythema

Axolotls' skin may be red, indicating stress, infection, or

injury. Monitoring for erythema allows for the early detection of potential health risks.

Filtration

Is the process of removing physical, chemical, and biological contaminants from aquarium water. Effective filtering is essential for both water quality and ecological health.

Floating Plants

Aquatic plants that float on the water's surface provide shade and reduce light intensity. Floating plants can also benefit water quality by absorbing surplus nutrients.

Gill filaments

The axolotl's external gills have delicate, feathery structures. These filaments facilitate gas exchange, allowing the axolotl to collect oxygen from the water.

Impaction

A condition in which an axolotl's digestive tract becomes clogged, usually owing to eating substrate or large food particles. Symptoms include bloating and loss of appetite. To avoid impaction, use a safe substrate and adequately sized food.

Infusoria

Tiny aquatic organisms are utilized to feed newly hatched axolotl larvae. Infusoria are simple to culture and provide vital nutrients throughout the early stages of development.

Mechanical Filtration

The physical removal of trash and particles from aquarium water by sponges, filter pads, or other media. Mechanical filtration helps to keep the water clean and free of waste.

Melanoid

Axolotl's color morph is distinguished by its dark, uniform

coloration and lack of iridophores. Melanoid axolotls can vary in color from dark brown to black.

Metamorphosis

Is the process by which some amphibians transition from larval to adult stages. Unlike many amphibians, axolotls do not undergo metamorphosis and retain their larval characteristics throughout their lives.

Nitrates (NO_3^-)

Beneficial bacteria convert nitrites into nitrates (NO_3^-), a less toxic consequence of the nitrogen cycle. While nitrates are less hazardous, they should still be kept low by changing the water often.

Nitrites (NO_2^-)

Are a harmful chemical created when helpful bacteria break down ammonia. Additional bacteria should convert nitrites to nitrates in a well-cycled tank and keep them at zero.

The pH level

Indicates the acidity or alkalinity of water. Axolotls prefer pH levels of 6.5 to 8.0. Maintaining a constant pH within this range is critical to their health.

P.V.C. Pipe

Polyvinyl chloride pipes are used in aquariums to build hides and tunnels for axolotls. P.V.C. pipes are strong, safe, and simple to adapt.

Regeneration

Axolotls can regrow lost or damaged body parts, such as limbs, tails, and even organs. This extraordinary characteristic is the focus of scientific inquiry and adds to their resilience.

Salinity

Is the concentration of salt in water. While axolotls are freshwater species, mild salinity increases (using aquarium salt) can be used to treat some health problems, such as fungal infections.

Sponge Filter

A sponge-based filter provides both mechanical and biological filtration. Sponge filters are mild and safe for axolotl aquariums because they do not generate high currents.

Substrate

The substance that lines the bottom of the tank. To reduce impaction and make cleaning easier, axolotls should be kept in a tank with fine sand or bare bottom.

Temperature Gradient

Temperature variations within the tank. Creating a temperature gradient can assist axolotls in locating their optimum thermal zone.

Thermometer

Measures the water temperature in a tank. Axolotl's health depends on maintaining the correct temperature, which a good thermometer may help.

Tubifex Worms

Axolotls commonly eat little aquatic worms. They are highly nutritious but must be sourced carefully to avoid introducing parasites.

Ultraviolet sterilizer

A device that employs U.V. light to destroy dangerous bacteria in water. U.V. sterilizers can help keep water clean and reduce the risk of sickness.

Water Conditioner

Chemical additions such as chlorine and heavy metals remove dangerous chemicals from tap water. Water conditioners are required to make tap water suitable for axolotls.

Water Hardness

A measurement of the concentration of minerals in water, specifically calcium and magnesium. Axolotls prefer somewhat hard water, which benefits their gill health and overall vitality.

Zygodactylous

Axolotls have two toes facing forward and two backward on each foot, which is a defining aspect of their limb shape.

This glossary of words is a thorough resource for learning the fundamental concepts and terminology involved in axolotl care. Whether new to raising axolotls or have had years of expertise, becoming acquainted with these words will help you give the best possible care for these unusual and fascinating creatures.

TROUBLESHOOTING GUIDE

Caring for axolotls can occasionally provide unforeseen complications. A troubleshooting guide will help you solve common problems and keep your pet healthy and happy. Let's look at some typical challenges and their remedies.

Water Quality Issues

One of the most common issues that axolotl owners confront is maintaining water quality. Poor water quality can cause a variety of health complications. Regularly test the water for ammonia, nitrites, nitrates, and pH. Ammonia and nitrites should always be nil, while nitrates should be less than 20 ppm. If you find ammonia or nitrites, make an immediate partial water change and inspect your filtration system. Before introducing your axolotl, ensure your tank is correctly cycled, and replace the water regularly to prevent harmful compounds from building up.

Temperature fluctuations

Axolotls thrive in chilly water, ideally between 60 and 68

degrees Fahrenheit. Higher temperatures induce stress and increase their susceptibility to sickness. If your tank is hot, consider moving it to a more relaxed place or installing an aquarium chiller. Fans blowing across the water's surface can also help reduce the temperature. If the water is icy, place your tank in a draft-free area and consider installing a heater built for small, gradual increases.

Appetite loss

When an axolotl refuses food, it might be alarming. First, look at the water parameters, as poor water quality is a common cause. If the water is safe, evaluate whether any recent changes in the surroundings could be generating stress. Axolotls may also go off food briefly while shedding or experiencing temperature fluctuations. To whet their hunger, offer a range of feeds, including earthworms, bloodworms, and high-quality pellets. If the resistance to feed persists, see a veterinarian with experience with amphibians.

Gills deteriorate

Gill deterioration may suggest low water quality, stress, or infection. Ensure that your water parameters are within the ideal range and that there is adequate oxygenation. Overcrowding can also cause stress and gill problems, so make sure your axolotl has sufficient space. If you suspect a bacterial or fungal infection, consult a veterinarian. To address the underlying cause, treatment options may include salt baths or prescribed drugs.

Floating

Floating can indicate digestive problems, such as gas or constipation, frequently induced by swallowing air when feeding at the surface or consuming substrate. If your axolotl is floating, determine whether it can return to the bottom. Offering living food that sinks, such as earthworms, can be beneficial. If the condition persists, gently massage the axolotl's abdomen to help evacuate trapped air and keep it from eating substrate using fine sand or a bare-bottom tank.

Injury and Regeneration

Axolotls can harm themselves on sharp items or while interacting with their tank mates. Ensure that all decorations and tank components are smooth and safe. If an injury occurs, keep the water clean to avoid infection and consider putting the axolotl in a separate tank to alleviate stress. Axolotls have excellent regeneration ability, but severe injuries must be adequately monitored. Adding a modest amount of aquarium salt can help encourage healing.

Unusual Swimming Patterns

Erratic or frenetic swimming may suggest stress, poor water quality, or the presence of an irritant. First, verify the water parameters and ensure the tank configuration is correct. Look for any sharp objects or aggressive tankmates that may be causing problems. Filter-generated strong water currents can also cause odd swimming behavior. Adjust the flow rate or deflect the stream to create a more peaceful environment.

Shedding

Shedding is a natural process. However, excessive or incomplete shedding might cause trouble. Ensure that your axolotl's diet is healthy and the water conditions are adequate. Poor water quality can disrupt the shedding process. Provide rough objects like driftwood to help your axolotl rub off the shedding skin. If the shedding persists, see a veterinarian.

Skin lesions or redness

Skin problems frequently signal bacterial or fungal infections, typically caused by poor water quality or trauma. Regularly monitor water parameters and keep the tank clean. If you see skin lesions or redness, isolate the axolotl in a separate tank and consult a veterinarian about appropriate treatments, including antibiotics or antifungal drugs.

Parasites

External parasites are uncommon but can cause substantial

discomfort. Signs include excessive rubbing against items, visible patches or worms, and lethargy. See your veterinarian for an accurate diagnosis and treatment plan if you suspect parasites. To prevent parasites from entering the tank, quarantine any new additions.

Understanding these frequent difficulties and their solutions can help you establish a healthy and stable habitat for your axolotl. Your axolotl will thrive if you observe it regularly, keep the water conditions appropriate, and proactively address concerns.

REFERENCES

1. Axolotls: Care and Breeding in Captivity by Peter W. Scott

   - A comprehensive guide covering various aspects of axolotl care, breeding, and maintenance in a home aquarium.

2. Axolotl: The Captive Care of Ambystoma Mexicanum by Stephen L. McGuirk

  - An informative book focusing on the specifics of axolotl care, habitat setup, and health management.

3. Amphibian Medicine and Captive Husbandry by Kevin M. Wright and Brent R. Whitaker

  - A detailed resource on the medical and husbandry needs of amphibians, including axolotls, written by veterinarians and experts.

4. The Biology of Amphibians by William E. Duellman and Linda Trueb

  - This book provides an in-depth look at amphibian biology, including physiology and behavior, which can help understand axolotl care.

5. Axolotl: Adventures in Mexican Periphery by Brian R. Sheen

- A more narrative-driven exploration of the axolotl, blending scientific information with personal experiences and stories.

6. The Axolotl: Zoology and Biology" by Gerhard Westermann

 - A detailed examination of the axolotl's anatomy, physiology, and unique biological traits.

www.ingramcontent.com/pod-product-compliance
Lightning Source LLC
Chambersburg PA
CBHW071912210526
45479CB00002B/384